ECONOMIC GEOGRAPHIES

Circuits, Flows and Spaces

RAY HUDSON

SAGE Publicatio.
London • Thousand Oaks • New Delhi

First published 2005

SAGE Publications Ltd
1 Oliver's Yard
55 City Road
London EC1Y 1SP

SAGE Publications Inc.
2455 Teller Road
Thousand Oaks, California 91320

SAGE Publications India Pvt Ltd
B-42, Panchsheel Enclave
Post Box 4109
New Delhi 110 017

British Library Cataloguing in Publication data

A catalogue record for this book is available from the British Library

ISBN 0 7619 4893 7
ISBN 0 7619 4894 5 (pbk)

Library of Congress Control Number 2004102659

Typeset by C&M Digitals (P) Ltd., Chennai, India
Printed in India at Gopsons Paper Ltd, Noida

Contents

Preface

During the 1990s, I worked intermittently on a book focused on geographies of production, eventually published in 2001 as *Producing Places*. As I did so, I was increasingly aware of three related developments within economic geography and more generally within the social sciences. First, there was an increasing interest on issues of consumption, complementing the previous emphasis on issues of production. Secondly, there was growing recognition of the grounding of the economy in nature and of the centrality of relations between economy and ecology. Thirdly, there was growing interest in cultural approaches to understanding economic geographies, which were seen as offering alternative 'bottom up' perspectives to the variety of 'top down' political economy perspectives that had dominated in the 1970s and 1980s (and which I had drawn on heavily in *Producing Places*). This raised a number of intriguing questions for me. They partly arose because my own research career as a doctoral student had begun in the late 1960s with an interest in how consumers made choices about shops and shopping centres, the knowledge that they had about the retailing environment, the criteria they used in making such choices, and with the processes through which they learned about new environments. They also arose because much or my subsequent work had investigated geographies of industries such as coal mining, steel making and chemicals production, each of which in different ways raised important questions about relations between economy and environment. More fundamentally, these developments revived questions as to understanding the totality of the production process in capitalism, and the significance of processes of exchange, sale and consumption in capital accumulation.

However, if only for reasons of space, there was no possibility of extending the scope of *Producing Places* adequately to consider these issues (although there was some consideration of them, especially economy/environment relations). Therefore, even while finishing work on that book, it seemed to me that there was a good case to consider writing another book that sought to develop an approach to economic geography that encompassed issues of production and consumption and that sought to develop a more nuanced pluri-theoretical approach. This would allow economic geographies to be explored in a number of registers and draw on political economy and cultural perspectives as complementary, while holding on to the strengths of political economy perspectives. Indeed, in many ways it sought to return to the traditions of political economy before the neo-classical marginalist revolution, reconnecting analyses of production and consumption, sensitive to both the cultural construction and material grounding of the economy in nature. This book is the result.

As always, it is at least as much a collective endeavour as it is an individual one, drawing on the generous help and assistance of numerous colleagues and friends. They include: Ash Amin; Kay Anderson; Huw Beynon; Costis Hadjimichalis; Peter Dicken; Nicky Gregson; Roger Lee; Gordon MacLeod; Doreen Massey; Linda McDowell; Stan Metcalfe; Jamie Peck; Helen Sampson; Ian Simmons; Denis Smith; Sue Smith; Adam Tickell; Nigel Thrift; Henry Yeung; Dina Vaiou; Paul Weaver; Sarah Whatmore and Allan Williams. In addition, several of my recent doctoral students have provided valuable ideas, especially Doug Lionais, Alison Scott, Leandro Sepulveda Amanda Smith, Delyse Springett and James Wadwell. Thanks also to Edward Arnold for permission to use a revised and edited paper from *Progress in Human Geography* as the basis for Chapter 1. As ever, however, the responsibility for the end product is mine.

Ray Hudson
Durham

1

Conceptualising Economies and Their Geographies

1.1 Introduction

In recent years there has been ongoing, at times heated, debate in economic geography as to how best to conceptualise and theorise economies and their geographies. During the 1970s and 1980s, stimulated by the critique of spatial science and views of the space-economy that drew heavily on neo-classical economics, strands of heterodox political-economy approaches in general and Marxian political economy in particular rose to prominence. These were important in introducing concerns with issues of evolution, institutions and the state, alongside those of agency and structure, in developing more powerful and nuanced understandings of economies and their geographies. Much of the subsequent debate in the 1990s was informed by post-structural critiques of such political-economy approaches, especially those that were seen (rightly or wrongly) to rely upon an overly deterministic and structural reading of the economy and its geographies (R. Hudson, 2001). These have been important in emphasising relationships between categories such as culture and economy and consumption and production and in provoking more serious consideration of issues such as relations between agency, practice and structure, between people, nature and things, the materiality of the economy and its discursive construction and representation.

Drawing on these recent debates, my primary focus is with exploring possibilities for developing more subtle and nuanced conceptualisations of economies and their geographies. This prompts two introductory questions. First, how do we best conceptualise the production of social life in general? Secondly, and more specifically, how do we most appropriately conceptualise 'the economy' and its geographies in capitalism? By 'the economy' I refer to those simultaneously discursive and material processes and practices of production, distribution and consumption, through which people seek to create wealth, prosperity and well-being and so construct economies; to circuits of production, circulation, realisation, appropriation and distribution of value. Value is *always* culturally constituted and defined. What counts as 'the economy' is, therefore, always cultural, constituted in and distributed over space, linked by flows of values, monies, things and people that conjoin a diverse heterogeneity of people and things. Equally importantly, the social processes that constitute the economy *always* involve biological, chemical or physical transformation via human labour of elements of the natural world. The resultant 'environmental footprint' of these activities emphasises the critical grounding of economies in nature. By 'capitalism' I refer to a particular mode of political-economic organisation defined by socially produced structural relations and parameters, which are always – and necessarily – realised in culturally and time/space specific forms. The extent to which the contemporary phase of capitalism represents a break from past trajectories of capitalist development continues to be a matter for debate.

While the prime focus of this book is the second introductory question, the conceptualisation of capitalist economies, it is framed by the first. Capitalist economies are constituted via a complex mix of social relations, of understandings, representations and interpretations, and practices. Certainly the class relations of capital are decisive in defining them as capitalist but these are (re)produced in varying ways and in relation to non-capitalist class relations and non-class social relationships of varying sorts (such as those of age, ethnicity, gender and territory). The social relationships of non-capitalist economies assume a great variety of forms, and occasional reference will be made to them. However, in order to allow some depth of analysis the focus will be on the economies of capitalisms and the social relations of capital that define and dominate them.

1.2   Six axioms and guiding principles and some of their implications

In seeking to answer the two introductory questions, I begin from six axioms. First, there is a need for concepts at varying levels of abstraction. This theoretical variety is necessary in order to describe and account for the diverse individual and collective practices, with varying temporalities and spatialities, involved in processes of production, distribution, exchange and consumption and in the spatio-temporal flows of materials, knowledge, people and value (variously defined) that constitute 'economies'[1]. All social life occurs in irreversible flows of time and has a necessary spatiality. Secondly, these diverse practices must be conceptualised as necessarily interrelated, avoiding fragmenting the economy into dislocated categories such as production and consumption, seen as at best unrelated, at worst hermetically sealed and self-contained. For a considerable period of time much social scientific analysis of the economy – whatever its theoretical stripe – tended to separate the analysis of consumption from that of production and, explicitly or implicitly, prioritised production over consumption. Consumption was simply seen as a necessary adjunct to production. Now this *is* the case in capitalist economies in one very precise sense. For both production and consumption – or, more accurately, exchange and sale – form moments in the totality of the production process. The point of sale is critical as it realises the surplus-value embodied in commodities and returns it to the monetary form. However, this is only a partial perspective on consumption. While services of necessity are (co)produced and consumed in the same time/space, the moment of sale of material commodities marks a shift in emphasis from their exchange to their use value characteristics, to what can be done with them post-sale in spaces of private and public consumption in homes and civil society. The post-sale life of commodities has important instrumental, material and symbolic connotations and dimensions (ranging from the creation of waste, to the giving of gifts based on relations of family, friendship, love and reciprocity, to the creation of identities).

Thirdly, knowledgeable and skilled subjects, motivated via various rationalities, undertake *all* forms of economic behaviour and practices. Although people are certainly not the all-knowing one-dimensional rational automatons of neo-classical economic theory, what they do, how they do it, and where they do it, are underpinned by knowledge and learning. People behave purposefully. They are active subjects, neither cultural dupes nor passive bearers of structures or habits, norms and routines. This touches on a key issue in understanding capitalist economies. For *the* key requirement of any form of capitalist production – the availability of labour-power – requires that people produced in non-commodity form become commodified, selling their capacity to work on the labour market in exchange for a wage. This requires understanding the processes whereby people are reproduced as sentient, thinking human beings, conscious agents with their own agendas, pathways and plans – that is, *not* as commodities – and the circumstances in, and the processes through, which they become commodified as labour-power.[2] Moreover, flows of people in the course of their actions within the economy (and in other arenas, such as those of family and community) can become a mechanism and medium for flows of knowledge. Such flows can occur both in the form of embodied knowledge (often tacit) and that of the transmission of information in codified forms (written, spoken) via a variety of media (letter, telephone, fax, e-mail, for example).

However, while the economy is performed and (re)produced via meaningful and intentional human action, knowledge does not translate in any simple one-to-one relationship to behaviour. Knowledge is a necessary but not sufficient condition. Action is much more than simply a product of information and knowledge, shaped by diverse influences, from emotion to economic possibilities. Thus because the performance of the economy always involves people, the spaces of the economy are always imbued with a variety of meanings beyond those of economic rationality. Moreover, people and organisations have differential abilities to acquire and use knowledge in pursuit of their various projects (although this is not to equate such behaviour with generalised self-reflexivity and the continuous monitoring of individuals' life projects: see Lash and Friedman, 1992). What people come to know and do depends in part upon their positionality in terms of class, ethnicity, gender and other dimensions of social differentiation and identity, and the powers and resources available to them by virtue of their position within the organisations and institutions of a given social structure.

Furthermore, intention does not translate in any simple one-to-one relationship to outcome. Purposeful behaviour may have unavoidable and unintended as well as, or instead of, intended outcomes (Miller, 2002, 166). This is because people chronically act in circumstances in which they lack complete knowledge of the context, of other people and objects, and of the relationships between the people and objects on which they act. There may be emergent properties because of the excess of practices, and the messy conjoining of people and things in heterogeneous networks and processes of ordering that produce emergence. Consequently, the unintended consequences of human action, from the individual to the formal organisations and institutions of the state, must be taken seriously (Habermas, 1976). Complex change may be unrelated to agents actually seeking to produce change. They may simply recurrently perform the same actions but 'through iteration *over time* they may generate unexpected, unpredictable and chaotic outcomes. Often the opposite of what human agents may be seeking to realise' (Urry, 2000b, 4). Nevertheless, given these qualifications about uncertainty, ignorance and unintended outcomes, an economy that is not underpinned by intentional, purposeful behaviour, knowledge and learning is simply, literally, inconceivable. Economic practices are performed by knowledgeable, socially constituted subjects, although the outcomes of their actions may differ from those intended, while the ways and forms in which knowledge and learning influence economic practice vary over space and time.

The fourth axiom follows from the third: the economy is socially constructed, socially embedded, instituted in a Polyanian sense (with institutions ranging from the informality of habits to the formal institutions of government and the state: Hodgson, 1988). These various institutions exhibit a degree of medium to long-term stability, set within the *longue durée* of structural parameters and necessary relationships that define a particular mode of political-economic organisation (such as capitalism). As such, the economy can be thought of as a relatively stable *social* system of production, exchange and consumption. However, institutional stability is always conditional and contingent – there are processes that seek to disrupt and break out of established institutional forms as well as processes that seek to reproduce them. Hollingsworth (2000, 624) emphasises that 'there is a great deal of path dependency to the way that institutions evolve'. Consequently, institutional evolution is path dependent, as economic practices are performed in and create real, irreversible time. However, this is also a conditional dependence, for there are forces that seek to break as well as reproduce path-dependency. Therefore it is more accurate to describe economic and institutional developmental trajectories as path-contingent, with periodic cyclical crises along a given path and the potential for secular changes from one path to another.

The fifth axiom is that institutionalised behaviour (individual and collective) is both enabled and constrained by structures, understood as stable yet temporary (albeit very long-term) settlements of social relationships in particular ways

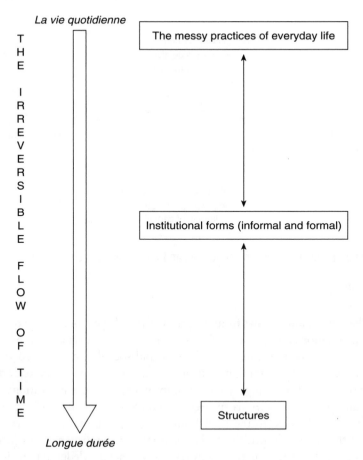

FIGURE 1.1 Temporalities of practies, institutions and structures

(see Figure 1.1). Structural relations specify the boundary conditions and parameters that define a particular mode of political-economic organisation as *that* mode. For example, the class structural relation between capital and labour is a defining feature of the capitalist mode of production – if this was not present, then some other mode of production would exist. However, this relationship can be constituted in varying instituted forms and this underlies the possibilities of creating many capitalisms and their historical geographies. Whatever the specific form, however, economic agents behave in instituted ways that are shaped by, and at the same time help reproduce, such structural relations. There are definite relationships between practices in the short term and in the long(er) term. Such relationships may be challenged – they often are. However, such challenges are typically folded into and absorbed in ways that alter, but do not radically break and transform, the defining structural characteristics and boundary conditions defined by capitalist social relations. Nevertheless, there is theoretical and, potentially, political space for structural change.

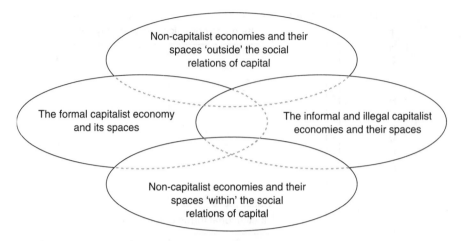

FIGURE 1.2   Relations between economies and their spaces in capitalism

The sixth axiom follows from the previous two. 'The economy' is constructed via social relations and practices that are not natural and typically are competitive. Consequently, they must be politically and socially (re)produced via regulatory and governance institutions that ensure the more or less smooth reproduction of economic life. These range from very informal governance institutions such as habits and routines in various spheres, including those of civil society, community, family and work, to the legal frameworks and formal regulatory mechanisms of the state. In short, the social relations of capital*ism* and not just those of *capital* must be reproduced (Figure 1.2), while acknowledging that the latter are both defining and dominant in capitalist economies and societies. However, while dominant, they are neither singular nor uncontested. Equally, there is a significant difference between the existence of rules and behavioural conformity with them. People may seek to contest or break rather than obey conventions and rules, raising key questions as to the circumstances in which they will do so. A distinction may be drawn between the formally regulated economy, the informal economy and the illegal economy. The formal economy consists of legal activities governed and regulated within the parameters of legislation. The informal economy consists of legal activities that are regulated by customary mechanisms and practices that fall outside the legal framework. Other activities are illegal but nonetheless form part of the economy (the economy of criminality, of the Mafia for example). However, the boundaries between formal, informal and illegal are fluid and vary over time/space.

The variety of institutions leads to complex spatialities of governance and regulation. These combine the diverse spaces and spatial scales of state organisations and institutions within civil society. Systems of governance and regulation are now more multi-scalar (Brenner et al., 2003) but national states retain a critical role within them (Sassen, 2003). While generally concerned with regulating the conditions that make markets possible, state activity can extend to supplementing or replacing market mechanisms in resource allocation, for example in the provision of welfare services or the production of key goods and materials. 'The economy' is chronically reproduced in situations of contested understandings, interests and practices only because governance and regulatory mechanisms keep such potential disputes within acceptable and workable limits. However, such mechanisms themselves must be socially (re)produced, often via processes of conflict and struggle. They do not simply emerge automatically to meet the functional needs of capital. Thus the practices of government, governance and governmentality are of critical importance. Furthermore, within forms of capitalism that encompass formal political democracies these mechanisms must be generally regarded as acceptable and legitimate but in dictatorial capitalisms they may be more violently enforced via coercive state power. One way or another, however, modes of governance and regulation must be sufficiently stabilised, at least for a time.

The requirement for a degree of admittedly contingent institutional – and even more so – structural stability reflects the need for a degree of predictability in the outcomes of economic practices and transactions. Economic actors – workers, banks, manufacturing companies and so on – require such predictability so that the transactions and practices of the economy can be performed with some certainty as to outcome over varying time horizons. Companies need to be confident that customers will pay their bills on time, workers that they will receive their wages regularly, and governments that tax revenues will arrive at the due date. Such stability is a necessary condition for a required degree of predictability as the capitalist economy is performative, a practical order that is constantly in action (Thrift, 1999). The requirement for stability is complicated precisely because of the dynamic character of the capitalist economy, the constant becoming of the economy. The economy is not something that simply is but always something that is necessarily in the process of becoming (as, for example, companies constantly strive to produce new things, in new ways). As such, there is an unavoidable tension between destabilising processes that would undermine predictability, stabilising processes that seek to assure it, and the necessarily dynamic character of capitalist production that complicates processes of governance and regulation and the smooth reproduction of capitalist economies.

In summary, the economy is instituted and structurally situated, produced by knowledgeable people behaving purposefully in pursuit of different and often competitive interests, which can be pursued with a sufficient degree of predictability of outcome and contained within acceptable – or at least tolerable – limits via diverse governance and regulatory mechanisms. There is an unavoidable tension between processes of institutionalisation that seek to create a degree of stability and predictability, and the emergent outcomes of practices that seek to disturb this, either deliberately or inadvertently. Consequently, no single totalising meta-narrative can explain everything about economies and their geographies but nevertheless meta-narratives remain valuable – indeed are necessary – in seeking such explanations (see Massey, 1995, 303–4).

More specifically, I argue that there are, broadly speaking, two analytic strategies for understanding the economy and its geographies, with different but complementary inflections. The first approach can be defined as '(political) economic', taking categories such as value, firms and markets as given, with these assumed to exist prior to their being observed and described from 'on high', and using them in analysing the economy. However, different types of economics conceptualise and represent these in different ways and I draw on Marxian and other heterodox traditions. The second can be thought of as 'cultural (economic)', with an epistemological focus on the discursive and practical construction and 'making-up' of these categories, while rejecting ontological claims that the economy has become more cultural. It emphasises the ways in which the economy is discursively as well as materially constructed, practised and performed, exploring the ways in which economic life is built up, made up, and assembled, from a range of disparate but always intensely cultural elements.

1.3   Neither a consumptionist nor a production be...

I want to enter a qualification at this point: any simple equation of production/ economic and consumption/cultural and of the primacy of the latter over the former, or vice versa, is illegitimate. While there is clearly a case for paying attention to consumption, there has been a tendency for the pendulum to swing too far, replacing one-sided productionist accounts with equally one-sided consumptionist accounts (Gregson, 1995), which, moreover, often conflate consumption with exchange and sale. This was especially so with 'first-wave' consumption studies. For example, Bauman (1992, 49) asserts that

> in present day society consumer conduct (consumer freedom geared to the consumer market) moves steadily into the position of, simultaneously, the cognitive and moral focus of life, the integrative bond of society ... in other words, it moves into the self-same position which in the past – during the 'modern' phase of capitalist society – was occupied by work.

Bauman alludes to the elision of (allegedly) post-productionist consumer society with post-modern society. Echoing this, Lash and Urry (1994, 296, emphasis in original) claim that 'the consumption of goods and services becomes *the* structural basis of western societies. And via the global media this [pleasure seeking] principle comes to be extended world-wide'.

There are two kinds of lesson to be drawn from this, which are reflected in 'second-wave' consumption studies. First, politically, it clearly exemplifies the dangers of confusing fashions in academic thought based on the class and socially specific experiences of an affluent minority with substantive changes in the living conditions and lifestyles of a much broader spectrum of the world's population. Even in the core territories of capitalism, only a small fraction of the consumption activities of the vast majority could be said to be 'pleasure seeking'. McRobbie (1997) criticises the political complacency of work on consumption that emphasises pleasure and desire precisely because it marginalises issues of poverty and social exclusion in its urge to reclaim the 'ordinary consumer' as a skilled and knowledgeable actor. For the vast majority of people living beyond the affluent core territories, hedonistic consumption and pleasure-seeking behaviour – let alone the attainment of pleasure – are distant pursuits of the affluent minority, occasionally glimpsed on television screens in a world characterised by perpetual hunger and malnutrition.

Secondly, theoretically, it illustrates the dangers of divorcing a concern with consumption from issues of production. Understanding capitalist economies and their geographies requires a more nuanced and subtle stance in theorising relations between the moments of consumption and production within the totality of the economic process and one that avoids equating the former with cultural-economy and the latter with political-economy approaches. There is, for example, a long history of rich ethnographic accounts that seek to understand work and the social relations of spaces of work in terms of the categories, understandings and practices of those engaged in the process (Beynon, 1973). Conversely, there are powerful political economies of consumption (Fine and Leopold, 1993; Miller, 1987). My argument is that such approaches are equally valid and should be seen as complementary. We need both to grasp the complexity of capitalist economies and their historical geographies, examining diverse practices of production, exchange and consumption from both political-economy and cultural-economy perspectives.

1.4 Cultural-economy approaches to understanding the economy

There has recently been a resurgence of emphasis on cultural approaches to understanding economies and their geographies, broadly falling into ontological and epistemological concepts of a cultural economy (Ray and Sayer, 1999). For example, Lash and Urry (1994) argue that there has recently been a significant 'culturalisation' of economic life, expressed in three ways. First, there has been a growth in the numbers of innovative companies producing cultural hardware and software. Secondly, 'there is a growing aestheticization or 'fashioning' of seemingly banal products whereby these are marketed to consumers in terms of particular clusters of meanings, often linked to lifestyles' (Lash and Urry, 1994, 7). Thirdly, there has been a growing 'turn to culture' in the worlds of business and organisations, precisely because maintaining or enhancing competitiveness requires companies to change the ways in which they conduct business and people to change the ways in which they behave within organisations.

However, the significance and validity of these epochal claims of increased culturalisation are far from assured. In certain limited respects the economy may have become more cultural but to claim that the economy overall has become more cultural is problematic. The evidence in support of 'the exemplary oppositions between a more "use"-value centred past and a more "sign-value-centred present"' is simply 'empirically insubstantial' (du Gay and Pryke, 2002, 7). Typically it is fragmentary, at times simply anecdotal. However, there is also an issue of adequate theorisation and conceptualisation of the links between economic and cultural. For, in practice, social actors cannot actually define a market or a competitor, '*except* through extensive forms of cultural knowledge' (Slater, 2002, 59, emphasis added). Producers cannot know what market they are in without considerable cultural calculation or understand the cultural form of their product and its use beyond a context of market competition. Understanding culture and (local) cultural difference is vital in order successfully to produce and sell globally (Franklin et al., 2000, 146). In like fashion, the economic practices of advertising, evocatively described as the 'magic system' (Williams, 1980), are intrinsically caught up with the cultural understanding of the role, functions and nature of advertisements (McFall, 2002, 161).

This draws attention to the way in which (to adopt a famous phrase from the cultural analysis of resources: Zimmerman, 1951): 'products and markets are not, they become'. This is perhaps most sharply emphasised by the iconic commodity of twentieth-century capitalism – the automobile – in which the cultural and economic are inextricably fused via market segmentation and the symbolic meanings associated with automobiles and automobility (Sheller and Urry, 2000). Furthermore, in order to be(come) a particular kind of economic institution, a market must also be a certain kind of culturally defined domain, dependent on the social categorisation of things as (dis)similar (Slater, 2002, 68). The dependence of markets upon such social categorisation undermines propositions about the increasing 'culturalisation of the economy' and the increasing, even complete, separation of the material and sign values of commodities.

Culturalism in its various forms reduces the product to its sign value and semiotic processes. As a result, the object becomes a dematerialised symbolic entity or sign, infinitely malleable and hence never stabilised as a socio-historical object; its definition can be entirely accounted for in terms of the manipulations of codes by skilled cultural actors. As such, the materiality of the object and the material economy and social structures through which it is elaborated as a meaningful entity are ignored. Consequently, there is also a tendency to reduce market structures and relations to semiotic ones. It is difficult to imagine how markets could exist over time, as they patently do, if products actually underwent the kind of semiotic reduction that culturalists assume. As Slater (2002, 73) notes, 'markets are in fact routinely institutionalised, and are even stabilised, around enduring definitions of products, whereas the semiotic reduction would assume that – as sign value – goods will be redefined at will'.

However, while the definition of a commodity, or of a thing, cannot be resolved by drifting off into the realm of floating signifiers, neither can its definition be simply and solidly anchored in given material properties. In contrast, the meanings of things, and things themselves, are stabilised or destabilised, negotiated or contested, within complex asymmetrical power relations and resource inequalities. This emphasises three things. First, the processes and interplay between the realms of the material, the symbolic and the social, through which the meanings of commodities are created, fixed and reworked. Secondly, the instituted social field within which multiple actors seek to intervene to establish the meaning of things. Thirdly, the political-economic structural relations within which actors and social fields are located.

Moreover, culture is located and performed in human and non-human material practices, which extend beyond human beings, subjects and their meanings, and implicate technical, architectural, geographical and corporeal arrangements (Law, 2002). Actant-Network Theorists such as Law and Latour (1987) conceptualise social production systems as heterogeneous networked associations of people and things and links among and between them. That the social has an irreducible materiality is – or ought to be – old news: 'Perhaps Marx told us this. Certainly Michel Foucault and a series of feminist and non-feminist partial successors have done so' (Law, 2002, 24). The reference to Marx is important, since one strand of the Marxian view of production centres on the labour process and transformation of elements of nature by people using artefacts and tools. In this regard, Law does no more – or less – than restate a proposition from Marxian analysis that conceptualises the economy as *always* a product of interactions between heterogeneous networks of people, nature (both animate subjects and inanimate objects) and things; of relationships between the social and the natural.

The conceptualisation of the economy therefore remains contested terrain, a terrain that is now more complex and in some ways more slippery in its analysis of relationships. Furthermore, this raises some important issues about the relationships between culture and economy. Miller (2002, 172–3) argues that it seems 'quite absurd' to suggest that we live within some new self-conscious, self-reflexive economy. There are undoubtedly powerful marketing discourses in the contemporary economy, but 'advertising and Hollywood were extraordinarily important' in the USA of half a century ago, and these made as much use as they could of current psychological theories about how to create subjects (Williams, 1980). On the other hand, the economy was just as cultural 'at the time when most academics saw themselves as Marxists'. It is undeniably true that a small, affluent minority live more self-reflexive (self-centred) lifestyles. It is also certainly an exaggeration to claim that there was a time when 'most' academics were Marxists. Neither point, however, negates the force of Miller's argument about the limitations of claims about the culturalisation of the economy.

In summary, positing a binary opposition between economy and culture is simply implausible and unhelpful. There is, however, considerable merit in an epistemological conception of cultural economy that envisages the cultural as a 'bottom-up' method of analysis,[3] complementary to a more top-down political economy. In contrast, suggestions that somehow the economy has (ontologically) become more cultural are misconceived. Miller (2002, 172–3) is particularly scathing in his comments about the 'culturalisation of the economy' thesis. He suggests that there seems to be 'a sleight of hand' through which a shift in academic emphasis is supposed to reflect a shift in the world, an economy that is more cultural than in earlier times. In this, he echoes Hall (1991, 20), who cogently argues that 'we suffer increasingly from a process of historical amnesia in which we think just because we are thinking about an idea it has only just started'. It is important to avoid such amnesia and to avoid conflating changes in the economy and changes in academic fashion. There is a need for eternal vigilance to guard against the constant danger of confusing new movements within thought (the (allegedly) new understanding that culture and economy cannot be theorised separately) from new empirical developments.

Classical political economy (as evidenced, for example, in the writings of Smith and Marx) prior to the marginalist revolution and the rise of neo-classical economics and its claims to universal economic laws recognised the cultural constitution of the economy (Amin and Thrift, 2004). For 'culture is everywhere and little has changed in this respect ... economically relevant activity has always been cultural' (Law, 2002, 21). Seeking to recover the ground conceded to the rise of neo-classicism in economics and acknowledge the long history of a cultural dimension within political economy is very different from assuming that there has been a qualitative change involving the culturalisation of the economy. The hard realities (if not quite iron laws) of commodity production and the production of surplus-value remain.

1.5 Political-economic approaches to understanding the economy

As with culture, the economic is a contested concept. There are several versions of the economic, based on differing theoretical presuppositions and forms and levels of abstraction. I reject technicist conceptions of the economy and its geographies exemplified by neo-classical and mainstream orthodox economics, which persistently seek methodologically to fix economic categories as self-evident or natural (and which are central to the (allegedly) new 'geographical economics': Krugman, 2000). Indeed, Slater (2002, 72) argues more generally that within economic analysis 'needs and goods appear as natural and self-evident'. In more critical theory, the use value/exchange value distinction within the commodity form 'has generally functioned as a proxy for the distinction for a "natural metabolism" between man and nature, and the warped social form taken by need and things within capitalist market relations'. While Slater's comments regarding neo-classical and mainstream orthodoxies are reasonable, his view of critical theory reveals a partial and warped understanding of Marxian political economy. For critical heterodox positions embrace more than Marxism while the notion of some unwarped natural form is difficult to reconcile with *any* notion of the economy as socially constituted and embedded.

Recognising the heterogeneity of heterodox economics, a political-economy approach needs to combine the differing but complementary forms and levels of abstraction of various heterodox positions – Marxian, 'old' and 'radical' institutional[4] and evolutionary. This multiple approach is needed to grasp the complexity of the economy as constituted by labour processes, processes of material transformation and processes of value creation and flow in specific time/space contexts. Marxian analyses allow specification of the structural features common to all capitalist economies that define them *as* capitalist. However, such structures do not exist independently of human practice; quite the contrary. They are both a condition for and an expression and result of such practices and are always contingently reproduced. Practices may give rise to emergent effects that challenge the reproduction of these structures, although there are powerful social forces and institutions that seek to assure their continuation. In short, there is a permanent tension between processes that seek to destabilise these structural relations and those that seek to reproduce them, which is generally – but not inevitably – resolved in favour of the latter. This may involve folding disruptive processes into new institutional forms of capitalism while leaving the defining class structural relations unchanged.

Indeed, the distinction within Marxian political economy between modes of production and social formations recognises that capitalism is constituted in variable ways. This insight has been considerably developed within other strands of heterodox political economy, in particular within evolutionary and institutional economics and sociology. Institutional approaches emphasise the ways in which economies are constituted and embedded in specific cultural and time/space contexts. Evolutionary approaches foreground the path-dependent character of development. At its most abstract level, the economy in capitalism is certainly dominated, indeed defined, by the social relations of capital – and powerful analytic tools are needed to theorise these. At this level, Marxian political economy and its value-theoretic account of the social relations and structures of capital provides powerful, highly abstract conceptual tools to understand accumulation by, through and as commodity production and surplus-value production. However, it is necessary to develop less (or differently) abstract concepts to understand how capitalist production and the (re)production of capital are secured and to capture the ways in which capitalism is instituted in specific time/space contexts, discursively and materially formed and concretised in and through specific informal and formal institutions. As such, it necessarily includes theorising the state, regulation and governance within capitalism and also links between the formal and informal sectors of capitalist economies (see McFall, 2002). Put another way, it requires understanding how practices, institutions and structures interrelate in the reproduction of capital (understood as a social relationship).

This in turn, however, requires acknowledging that the commodity form within capitalism is a slippery one, temporally and spatially (Appadurai, 1986), and that the social structural relations of capital intersect and interact in 'structural conjunctions' (Miller, 2002, 166) with those of other social structures (such as ethnicity or gender). While there may be co-evolution of structures, this is a variable and contingent process. Massey (1995, 303–4) recognises that there are broad social structural relations – of class, gender and ethnicity, for example, which have determinate though non-deterministic effects. Recognition of such broad structures 'is not the same as the commitment to, or the adoption of, a meta-narrative view of history. None of the structures ... need to be assumed to have any inexorability in their unfolding ... outcomes are always uncertain, history and geography have to be made'. These effects are determinate rather than deterministic precisely because of the multiplicity of structures, the conjunctural specificities of which combination of structures intersect and interact in a given time/space (which may also activate specific local contingencies) and the emergent properties of practices.

The process of commodification brings about, albeit unevenly, the extension and penetration of capitalist mechanisms and forms into aspects of the world and lifeworld from which they were previously absent. However, these processes result in uncertainty about the fate of commodities once they have been sold. The purchase of commodities depends (*inter alia*) upon the meanings that consumers attach to them. Consumption is one source of meaning and identity, both for those purchasing the commodity and those consuming it. However, such identity values are subject to change and renegotiation. Not least this is because commodities are manufactured with their own pre-planned trajectories, with built-in obsolescence within a product life cycle. As commodities reach the end of their socially useful lives to their original purchasers, they may be re-sold, both formally and informally in a variety of spaces (such as street markets and car boot sales). In this process, the meanings attached to commodities by their original purchasers are typically reworked (as, often, are the things themselves) so that there are recursive circuits of things and meanings rather than simply linear paths or a single circuit of meaning.

Capitalist economies include economic activities that are not under the direct sway of capitalist relations of production, both within and outside the spaces of capitalism (Figure 1.2). Conceptualising how capitalist and non-capitalist economies relate to one another and understanding strategies of 'accumulation by dispossession' – that is, (forcibly) taking things/people not produced as commodities and commodifying them (Harvey, 2002) – become important issues.[5] Conceptualising the economy therefore requires recognising the existence of non-capitalist social relations within capitalist economies and non-capitalist economies alongside capitalist ones and considering different concepts and theories of value and other economic categories from those appropriate to the mainstream, formal capitalist economy. It requires consideration of different processes of valuation, in which value is not defined as socially necessary labour time but in terms of some other metric, perhaps in a more multidimensional way that reflects a broader range of cultural and social concerns. It also requires consideration of processes of production and consumption in these alternative economies and their circuits, flows and spaces and of their (lack of) relationships to the mainstream (Leyshon et al., 2003). This raises questions of the political character of political economy and leads into a normative question of future alternatives, of 'sustainable economies' and their spaces.

1.6   Cultural economy and political economy: complementary approaches

While some see cultural economy and political economy as alternative approaches, I prefer to see them as complementary perspectives: understanding geographies of economies needs to embrace both. This does no more than recover a position that was central to classical political economy but that was generally (there were exceptions) denied for many decades following the ascendancy of neo-classical orthodoxy and that continues to be denied within the discipline of mainstream economics. Nevertheless, such recovery is vital to a more nuanced understanding of economies and their geographies. Thus objects of analysis can be both taken as given and problematised in terms of their discursive and material constitution. Du Gay and Pryke (2002, 2) suggest that 'the turn to culture' reversed the perception that markets exist prior to, and hence independently of, descriptions of them. A cultural approach indicates the ways in which objects are constituted through the discourses used to describe and act upon them. As such, economic discourses format and frame markets and economic and organisational relations, ' "making them up" rather than simply observing and describing them from a God's-eye vantage point'. This has critical analytical implications since it suggests that 'economic discourse is a form of representational and technological (that is, cultural) practice that constitutes the spaces within which economic action is formatted and framed'. Put slightly differently, the discursive space of the economic decisively shapes the practical spaces of the economy, and vice versa. Discursive and practical spaces are co-determining, co-evolutionary.

As such, economic categories (for instance firms, or markets) need to be analysed in complementary ways that acknowledge the processes through which commodities are produced and the meanings of commodities are created, fixed and reworked and the political-economic structural relations in which people are unavoidably located. What is required, therefore, is a culturally sensitive political economy that begins from the assumption that the economy is – necessarily – always cultural and a politically sensitive cultural economy that is alert to the power geometries and dynamics of political economy. These provide complementary approaches, viewing the economy from different analytic windows rather than an either/or ontological and epistemological choice. Indeed, these approaches in some respects interpellate one another rather than being discrete and self-contained. As such, the space currently occupied by culture-economy divisions and reductions could be at least partially reconstructed by treating concepts such as competition, markets, products and firms as *both* lived realities *and* as formal categories (see Slater, 2002, 76). Moreover, it could reasonably be argued that Marxian political economy has always contained strands of both approaches (Anderson, 1984).

1.7 Reconsidering the issues

Given the above, I now want further to explore two sets of interrelated issues. First, the conceptualisation of relations between agents, practices, representations and structures and their varying temporalities (Figure 1.1), using the notion of practice as what people do in the economy as a way of better grasping relationships between agency and structure by emphasising doing rather than just thinking, the material and affective as well as the cognitive. Practices are 'materially heterogeneous relations' that 'carry out and enact complex interferences between orders or discourses' (Law, 2002, 21–3). As such, diverse economic practices interfere in different and specific performances with other, alternative strategies and styles, producing an 'irreducible excess' which is necessary to the survival of discourses and performances grounded in them. The second set of issues centres on the conceptualisation of relations between spaces, flows and circuits, addressing the question of how to explain which parts of circuits are fixed in which spaces for a given period of time. Three points can be made briefly. First, spaces must be understood relationally, as socially constructed. Secondly, economic process must be conceptualised in terms of a complex circuitry with a multiplicity of linkages and feedback loops rather than just simple circuits or, even worse, linear flows (see Jackson, 2002).[6] Thirdly, the economy must be conceptualised as a complex system, a *fortiori* given recognition that it involves material transformations and co-evolution between natural and social systems.

There are two important implications of complexity in this context. First, economic practices may have unintended as well as, or instead of, intended consequences, because people chronically act in circumstances of partial knowledge. Secondly, complexity implies emergent properties that *may* lead to change between developmental trajectories rather than simply path-dependent development along an existing trajectory. There is a danger that concepts of path-dependency (especially if grounded in biological analogy) can lead to an underestimation of the role of agency and reduce actors to 'cultural dupes' (Jessop, 2001). People thus cease to be knowledgeable actors and come to be regarded as the passive bearers of habits, norms and routines (much as structuralist readings of Marx reduced them to passive bearers of structures). As a result, the concept of path contingency better expresses the possibilities of moving between as well as along developmental paths (Hardy, 2002). Actions and practices and systemic interactions may create emergent properties that alter, incrementally or radically, the direction of developmental trajectories. Consequently, evolutionary paths may be far from straightforward. As such, recognition of complexity and emergent properties can aid understanding of a shift from a simple evolutionary perspective of change along a given trajectory to evolution understood as a change from one trajectory to another. However, it is an open question as to whether emergent properties lead to changes within the parameters of capitalist social relations or to a shift to alternative non-capitalist paths.

There has been a lively – at times, heated – debate as to the conceptualisation of contemporary economy and society in terms of circuits, flows or spaces, and of the relations between them. Some argue that fixities no longer matter, or matter less, in a world of flows and (hyper)mobilities (Castells, 1996; Urry, 2000a). There is undeniably evidence of greater mobility, albeit unevenly, across a wide range of activities and spatial scales.[7] But for social life to be possible, for the economy to be performable, fluid socio-spatial relations require a degree of permanence, of fixity of form and identity – whether in terms of the boundaries of firms, national states or local spaces.

However, there is also a dialectic of spaces and flows and circuits, centred on the *necessary* interrelations of mobilities and fixities. Circuits and flows require spaces in which their various stages/phases can be performed and practised, while stretching social relations to create spaces of different sorts, fixing capital in specific time/space forms and ensembles (R. Hudson, 2001, Chapter 8). Spaces are both discursive and material. Material spaces are constituted as built environmental forms, a product of materialised human labour. Discursive spaces enable meanings to be both contested and established, permissible forms of action to be defined and sanctioned, and inadmissible behaviour to be disciplined. Recognising that spaces are discursively constructed implies that this process does not simply describe the economy. It is also, in part, constitutive of it, defining the economy as an object of analysis, constructing the spaces of meaning and the meaning of the spaces in which the economy is enacted and performed. These spaces of meaning then become guides to social and individual action. The same point can be made about concepts of circuits and flows, which are also constitutive rather than simply descriptive. As such, spaces, flows and circuits are socially constructed, temporarily stabilised in time/space by the social glue of norms and rules, and both enable and constrain different forms of behaviour.

Spaces, flows and circuits are thus both the medium and products of instituted practices (over varying time scales), based on human understandings and knowledges, situated in specific time/space contexts. As such, they are socially constructed and shaped (but not mechanistically determined) by prevailing rules, norms, expectations and habits, and by dominant power relations. As Law (2002, 24) remarks of factories, markets, offices and other spaces of the economy, each is 'a set of socio-technologies and a set of practices. But socially it is also a set of rules.' Such spaces are thus simultaneously materially constructed, a fixation of value in built form, a product of and an arena for practices, defined and regulated by socially sanctioned rules which prescribe or proscribe particular forms of behaviour. In this sense there are structural limitations on action and understanding but, reciprocally, these limitations are a product of human action, beliefs and values: structures are both constraining and enabling. Structural constraints are most powerful when they are hegemonic, taking effect because they have become taken-for-granted, unquestioned determinants of everyday behaviour (Gramsci, 1971). Everyday routine then – even if unintentionally and unconsciously – reproduces these structural relations. 'Enabling myths' (Dugger, 2000), deeply embedded in the beliefs and meanings in which such routine is grounded, have the effect of 'naturalising' the social and reproducing the structural. However, as structures do not exist independently of human action and understanding but are always immanent, contingently reproduced, they are in principle changeable. This is a key theoretical point and – potentially – one of great political importance.

Bourdieu catches this sense of hegemony via his concept of habitus. Habitus emphasises the doxic (taken-for-granted, unthinking) elements of action, social classification and practical consciousness. He (1977, 72) argues that the structure of a particular constitutive environment produces 'habitus, systems of durable, transposable dispositions, as structured structures, that is, as principles of the generation of practices and representations which can be objectively regulated and "regular" without in any way being the product of obedience to rules'. They are 'objectively adapted to their goals without pre-supposing a consensus aiming at ends or an express mastery of the operation necessary to attain them, and being all this collectively orchestrated without being the product of the orchestrating actor of the conductor'. Bourdieu (1981, 309) later makes a critical point in insisting that habitus is 'an analytic construct, a system of "regulated improvisation", or generative rules that represents the (cognitive, affective and evaluative) internalisation by actors of past experience on the basis of shared typifications of social categories, experienced phenomenally as "people like us" that varies by and is differentiated between social groups. Crucially, however, 'because of common histories, members of each "class fraction" share similar habitus, creating regularities of thought, aspirations, dispositions, patterns of action that are linked to the position that persons occupy in the social structure they continually reproduce'. While Bourdieu refers specifically to 'class fractions', commonality of experience and identity could equally be based on 'people like us' defined via other social attributes, such as ethnicity, gender or place of residence. Furthermore, historical processes of class formation will reflect the intersection of structures of class relations with those of other social structures (see Massey, 1995, 301–5).

1.8   Taking stock

We need to take what people do and their reasons for doing it, their actions and performances, seriously if we are to understand how structures are (un)intentionally (re)produced and constitute guides to action, informing social agents of appropriate ways of going on. For example, capitalists and workers behave in particular ways because they understand the world in terms of a specific class structural representation of capital/labour relations. Nationalists and regionalists behave in particular ways because of their understanding of the world as principally organised around shared territorial interests and identities. Moreover, such behaviour may well be paradoxical precisely because social actors behave in circumstances beyond their control. For example, radical tradeunionists go to work, even though they understand the capitalist labour process as exploitative, since on a quotidienne basis they and their families need to eat, to have a place to live and so on.

The economy is thus instituted, discursively established based on shared understandings regarding proper behaviour and conduct by the owners and managers of capital and the vast variety of workers in factories, offices and shops, by the mass of consumers and so on. But these shared understandings and resultant practices/performances are structured by the *necessary* requirements of capitalist production: a sufficient mass and rate of profit. As such, they are shaped by and simultaneously help reproduce structural constraints and the materiality of the economy. Thus, capitalist business is based in a material culture of relations between people and things that ranges from the vast number of intermediaries required to produce trade, through the wide range of means of recording and summarising business, to the different arrangements of buildings (spaces of work) that discipline workers' bodies. These devices and arrangements 'are not an aid to capitalism; they are a fundamental part of what capitalism is' (Thrift, 1999, 59). Indeed, many of them are produced as commodities.

The recognition of different arrangements of buildings as spaces of production that discipline workers' bodies touches on an important aspect of the ways in which spaces of economies are both a medium for and product of human behaviour. More generally, economic spaces, circuits and flows both help produce and are (re)produced by performance. They both constrain and enable different forms of economic practice. In this way consumers and producers of these spaces both produce and consume their own (formally economic) citizenship. Those who cannot produce or consume in this way cease to be legitimate citizens. Spaces and practices are 'binding agents' in terms of how economies are performed and subject positions created and inflected (Thrift, 2000); the same point can be made about circuits and flows. Alternatively, and simultaneously, they are agents of social exclusion for those denied access to them.

However, relationships between agency, practice and structure are further complicated because (as the Foucauldian comment about disciplining workers' bodies hints) there are typically contested and competing understandings of what is and what is possible in terms of action and change. For example, there is a struggle within spaces of work between managers and workers to define and dominate the 'frontier of control' (Beynon, 1973). Equally, there are typically competitive struggles between capitalists for markets and profits and among groups of workers seeking to promote their interests in competition with other groups of workers (R. Hudson, 2001). All must also be disciplined to accept the rules of the game of the commodity-producing market economy in conducting these struggles, though these rules vary through time/space. As a result, there is a complicated and multidimensional struggle for domination between competing views of the world and material interests. Consequently, the reproduction of structural constraints is a product of contested processes, unless, of course, one particular view becomes generally if not universally accepted as hegemonic.

1.9   A map of the remainder of the book

The organisation of the remainder of the book reflects this exploration of the constitution of and relations between spaces, circuits and flows in capitalist economies. It also reflects a view of space and flow as providing complementary analytic lenses through which the 'same' event or phenomenon can be viewed, depending upon the purpose of analysis. Thus Chapters 2–5 primarily focus on flows of capital, materials, knowledge and people, and Chapters 6–9 on spaces of governance, production, sale and consumption. Chapter 10 addresses the issue of sustainable spaces and flows.

Notes

1 Gough (2003) castigates me for not rigorously deducing such concepts from the value categories of capital (see R. Hudson, 2001). But to do so would be to seek a single totalising meta-narrative account that can explain anything and everything.
2 In some circumstances slavery and indentured labour have become mechanisms to assure the supply of labour-power.
3 Methodologically, this involves ethnographies, participant-observation and interviews, well-established approaches in the social sciences and in seeking to understand economic forms and practices.
4 'Old' institutionalism emphasises the institutional and social embeddedness of the economy. Radical institutionalism emphasises asymmetrical power relations in shaping economic life. In contrast, 'new' institutionalism is close to mainstream orthodoxies (Hodgson, 1993).
5 See also Amin (1977) on the articulation of modes of production.
6 Such as commodity chains (Clancy, 1998).
7 Damette (1980) introduced the concept of hypermobility of capital. Thus the notion that the capitalist economy has suddenly 'speeded up' requires careful consideration.

2 Flows of Value, Circuits of Capital and Social Reproduction

2.1 Introduction

At a rather high level of abstraction and generalisation, all forms of economy and society may be conceptualised as reproduced via continuous flows of value as products circulate between people, times and spaces. These flows of value through the sequence of production, exchange and consumption are both constituted in and help constitute circuits of social reproduction. Value is generated through relations and things which, via the material and social practices of the economy, come to be socially regarded as useful, helpful, uplifting or, more narrowly but generally, as fundamental to life going on 'as normal'. These flows encompass the exchange of value embodied in products and may involve the exchange of money for work or the capacity to work, which could increase future production and/or consumption. Social and material survival requires that circuits of social reproduction deliver such flows of value, in appropriate quantities, distributions and time/spaces. In turn, successfully maintaining such circuits involves complex intersections of material and social relationships and practices in the formation and definition of value. The material and the social are intimately related via circuits of co-evolution and co-determination: 'the significance of any single moment of economic activity begins to make sense in material terms only in the context of circuits of material reproduction' (Lee, 2002, 336). Material relations, imbued with social meaning, involve the practice and co-ordination of circuits of production, exchange and consumption. Thus social relations define the meanings of material practices. These relationships and meanings become voluntarily, often unquestioningly, accepted and confer a sense of social order via the recurrent practices of the economy. In other circumstances, maintaining social reproduction requires deployment of coercive power within circuits of authority, control and direction to shape economic processes and circuits of material practices.

The substantive content and meaning of conceptions of value are spatio-temporally specific, related to different modes of socio-economic organisation and different ways of conceptualising these and theorising 'the economy'. For example, in the formal capitalist economy value can be defined as market price (as in neo-classical and mainstream orthodox theories), or as socially necessary labour time (as in the orthodox Marxian labour theory of value or as in the value theory of labour: Elson, 1979). Thus in its conceptualisation of flows of value and circuits of capital in the formal capitalist economy, Marxian analyses recognise that values and prices are not synonymous and that the relationships between them need to be considered. Not least this is because of the characteristics of money and monetary systems and the disciplining power of money on the practices and developmental trajectories of economies. Beyond the confines of the formal economy and theorisations of it, there are alternative conceptions of value in terms of labour time (as in Local Exchange Trading Systems, or LETS), or in terms of the intrinsic worth of things. Such alternative conceptions of value influence economic practices within the spaces of economies of capitalism, permeating the interstices that capital has abandoned or never found sufficiently attractive, or from which it is prohibited by regulation or morality, custom and the force of tradition. As such, it emphasises that capitalist economies involve complex and multiple flows of values, and that different conceptions of value may underlie these flows. Such conceptions vary with the form and type of economic organisations and the positionality of those constructing the category of value.

2.2 The production of value and three forms of the primary circuit of capital

Capitalist relations of production dominate the contemporary world. Mainstream capitalist production encompasses value expansion via the production and realisation of surplus-value: it is simultaneously a labour process and a process of valorisation. This form of organising production is based on the class structural separation of, and dialectical relation between, capital and labour. These social relations are extended and stretched over space and imposed upon people to define the character of social reproduction – although not without contestation and resistance – by commercial and financial institutions operating over multiple spatial scales from the very local to the global. Such institutions define highly focused notions of value directed at profitability and accumulation and use them to constrain and direct capitalist circuits of social reproduction – that is, the expansion of capital and the extension and deepening of capitalist social relationships. In this way, the social and material dimensions of reproduction are mutually formative and inseparable and take a specific form within the parameters of capitalist social relationships.

The emergence of capitalist production and associated flows of value to a position of dominance has been, and is, temporally and spatially specific. As production evolved historically, there was a gradual shift towards the creation of a social surplus beyond the immediate needs of producers. This enabled the transition to systems of production for exchange, which provided crucial preconditions for the subsequent emergence of capitalist relations of production. First, the permanent production of a surplus and the development of a social division of labour provided the necessary economic conditions to allow (but not determine or guarantee) the emergence of social classes. Secondly, the development of money as a specific commodity to facilitate exchange was critical (Smith, 1984, 35–47) because the use of money as an individualised and exclusionary form of social power is a central feature of capitalism (Harvey, 1996, 236). Thirdly, the transition to production for exchange necessarily involved the alienation of both consumer and producer from the product, a critical move in the creation of markets and patterns of consumption and in the organisation of the labour process. Not all production in the contemporary world is for exchange, however, and not all production for exchange is capitalist.

One way to clarify what is specific about *capitalist* commodity production and identify its distinguishing features is to draw upon the Marxian concept of mode of production, a quite abstract conception of social and economic structure. This characterises particular types of economic organisation in terms of specific combinations of forces of production (artefacts, machinery and 'hard' technologies, tools – in short, the means of production) and social relations of production. In the capitalist mode of production *the* key defining social relationship is the class structural one between capital and wage labour. This is a dialectical and necessary (in the critical realist sense: Sayer, 1984) relationship because capital and labour are mutually defining; the existence of one presupposes the existence of the other. Since living labour is the only source of new value (surplus-value) created in production, capital needs to purchase labour-power in order to set production in motion. Conversely, labour needs to sell its labour-power for a wage in order to survive and reproduce itself. The specific and distinguishing characteristic of the capitalist mode of production is that it is structured around the wage relation, with labour-power bought and sold in a market like any other commodity. The key point, however, is that labour-power is *not* like any other commodity – and this is critical in understanding capitalist economies and their geographies.

Commodities simultaneously possess attributes as use values and as exchange values. As materialised human labour, transformed elements of nature, they have qualities that people find useful and, as such, use values. These use value characteristics reflect the concrete aspects of labour, the fact that labour is private and specific. At the same time, labour within capitalist relations of production is also abstract labour, universal, social and general in so far as it defines the exchange values of commodities on world markets (Postone, 1996). Abstract labour is 'a remarkable thing', simultaneously a social relationship, a measure of value, a determinate magnitude (socially necessary labour time), causally efficacious and invisible yet real (Castree, 1999, 149). In the capitalist mode of production the exchange value of a commodity is defined as the quantity of socially necessary labour time – that is, the amount of undifferentiated abstract labour needed under average social and technical conditions of production – required

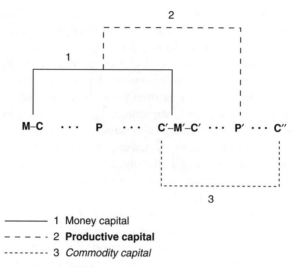

FIGURE 2.1 Exchange sequences

to produce it. Abstract labour and socially necessary labour time are therefore central to understanding the rationale of capitalist production and associated flows of value. For the driving rationale of a capitalist economy is production for exchange and profitable sale through markets; that is, the production of exchange values in contrast to, say, self-sufficiency, with each person or household producing all that (s)he or it needs or to the maximisation of physical output *per se*. Capitalist production is thus organised with the *purpose* of sale in markets, in contrast to the sale of a fortuitously produced surplus in a subsistence economy. Production therefore finds its rationale in, and is socially validated *ex post* by, market exchange and the successful sale of goods and services rather than *ex ante* by state planning or via some other criterion.

Understanding how qualitatively different use values become exchanged as quantitatively equivalent exchange values requires some consideration of exchange sequences within the primary circuit of capital, and in particular in that of productive industrial capital (Figure 2.1). In the exchange sequence C–M–C' a given amount of money is used to purchase one commodity – for example, a pair of shoes – and the seller of the first commodity then uses that money to buy another and different commodity – say a radio. Any such C–M–C' sequence implies exchange for money and resultant flows of material commodities between producers and consumers, between firms, and within and between spaces, as registered in inter-regional and international trade, for example. Money therefore functions as a medium of exchange, allowing the quantitatively equal exchange of two qualitatively different commodities. Equally, consider a sequence M–C–M'–C' where C'>C and M'>M. This could describe the moment of exchange and the activity of retailing, 'the almost mystical transformation of commodity capital to money capital and back to commodity capital' (Blomley, 1996, 243). In this case, the retailer advances money capital to purchase commodities for re-sale and receives a portion of surplus-value as a necessary condition of maintaining the circuit of value and continuity of the accumulation process by virtue of the difference between the buying and selling prices of commodities. Part of the realised money capital is in turn used to purchase further commodities for re-sale and the exchange sequence continues. Now consider a rather different sequence, C–M–C'–M' in which again C'>C and M'>M. For example, companies purchase commodities produced by other companies that then form inputs to their production process. Clearly, in this case money is not simply functioning as a medium of exchange or as money capital that receives a share of surplus-value by virtue of performing the necessary work of exchange and realisation of already created surplus-value. In fact, in this case money capital is being advanced to purchase commodities that are transformed into commodities of greater value, to be sold in order to make more money, to make profits. This therefore raises a seminal question: where does this profit come from? It cannot, in a systematic and systemic sense, originate in the process of circulation, precisely because the exchange process involves the exchange of equivalents via market transactions carried out between formally free and equal agents. It can then only originate in the process of production itself and the way that this is structured within capitalist relations of production so that the value of commodities purchased as inputs to production is exceeded by the value of commodities resulting at the end of production – hence Marx's emphasis on the centrality of this moment in the totality of the production process.

Capitalist production can therefore be usefully thought of as a continuous circuit, with the primary circuit encompassing three analytically distinct yet integrally linked circuits: commodity capital; money capital; productive industrial capital (Figure 2.2). Clearly, such circuits have definite geographies, with different locations forming sites of production and exchange, linked by flows of capital in the forms of money, commodities and labour-power. Over time, the spatial reach of such circuits has increased, with the circuits of commodity, money and productive capital successively becoming internationalised (Palloix, 1977). The circuit of productive industrial capital provides key insights to understanding the creation and realisation of surplus-value, of profits and the dynamism of geographies of the production process in its totality within the social relations of capital. This circuit requires that capital be first laid out in money form to purchase the necessary means of production (elements of constant and fixed capital in the forms of factories and buildings,[1] tools, machinery, manufactured inputs and raw materials) and labour-power.[2] Labour-power and the means of production are then brought together in the production process, in the workplace, under the supervision of the owners of capital or their managers and representatives. Two things happen in the moment of production. First, existing use values, in the form of raw materials, machinery and manufactured components, suitably revalued according to their current cost of production, are

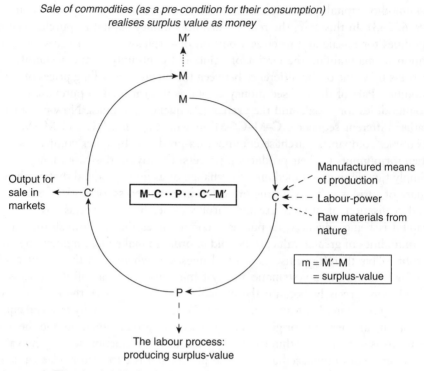

FIGURE 2.2 The circuit of industrial capital

transferred to new commodities. Secondly, surplus-value is created. This augmentation of value is possible precisely because labour-power is a unique fictitious commodity. For capital purchases not a fixed quantity of labour but rather the workers' capacity to work for a given period of time. In this time, workers create commodities that embody more value than was contained in the money capital used to purchase their labour time. This difference in value is the surplus-value, the additional new value created in production, which is realised in money form as profits on successful sale of the commodity, along with existing values transferred in the production process. In brief, capitalist production is simultaneously a labour process, producing material use values, and a valorisation process, reproducing value and producing surplus-value, which is embodied in commodities and, having been realised, flows through the economy.

   The smooth flow of capital around the circuit is thus necessarily interrupted as capital is fixed and materialised in specific commodified forms (automobiles, houses, shirts and so on). In some cases the value and surplus-value that these commodities embody can be realised quite quickly and capital thrown back into circulation. In others, however, the process of amortisation can take years, even decades, as capital is fixed in built forms of great durability and duration. Moreover, realisation is by no means guaranteed for *any* commodity, Capitalist production is an inherently speculative and risky process, with a constant danger that the circuit might be broken or interrupted in non-renewable ways. Assuming that sale is successful, however, the difference between the amount of money capital advanced at the start of the round of production and that realised at the end of it is equivalent to the difference in the value of commodities at the beginning and the end of the round. This is critical in understanding the rationale and dynamism of capitalist production. It also emphasises that the totality of production involves more than simply the transformation of materials to produce goods or services. It also involves a myriad of other activities associated with transportation, distribution and sale, since the determination of socially necessary labour time is contingent upon 'socially necessary turnover time', the speed with which commodities can be distributed through and across space (Harvey, 1985). Furthermore, the final consumption of goods and services and the meanings with which they are endowed, the identities that they help create and form, are of central importance as final consumers purchase commodities in the belief that they will be useful to them, materially and symbolically.

In summary, the circuit of productive industrial capital conceptualises commodity production, exchange and consumption in terms of the creation, realisation and flow of value. It also emphasises that commodity production is inherently geographical in a double sense. First, the material transformation of natural materials is predicated on relationships between people and nature: a social–natural dialectic. Secondly, space is integral to the biography of commodities, which move between varied sites of production, exhange and consumption as they flow around the circuit: a socio-spatial dialectic. The circuit of productive capital thus involves complex relationships between people, nature and space in processes of value creation and realisation and in flows of value through time/space. Conceptualising production as a process of successive journeys around the circuit of industrial capital aids understanding of developmental trajectories within capitalism, both at the level of the individual firm and of capital in general. In particular, it helps reveal what happens to the money equivalent of the newly produced surplus-value.

Two limit cases can be established: simple reproduction and expanded reproduction of capital. First, in the case of simple reproduction, the next round of production would begin with the advance of the same amount of capital as the previous one because the entirety of surplus-value produced had been conspicuously consumed on luxuries by the capitalist class. Secondly, with expanded reproduction, it is used to increase the scale of production, with all the surplus-value thrown into circulation as money capital. This corresponds to the maximum possible rate of growth. In practice, the outcome typically lies between these two extremes. There is an augmentation of capital but this is usually less than the maximum feasible amount of capital that could be advanced. The temporal fluctuations in this amount help define the cyclical variations of the business cycle around a longer-term secular growth trend of expanded accumulation.

The conditions necessary for sustainable (economically as opposed to ecologically) expanded reproduction and the smooth and uninterrupted flow of value have different implications depending upon whether the focus is upon an individual company or capitalist production overall (Mattick, 1971). An individual company is subject to contradictory pressures. It simultaneously wishes to minimise its own input costs and to maximise its sales and profits. Maximisation of the latter depends, however, upon purchases by other companies (seeking to minimise their costs) and final consumers (whose wages may represent a significant proportion of other firms' costs of production). As such, a capitalist economy is reproduced via contradictory processes; it travels along an uncertain and crisis-prone trajectory, with an immanent tendency to interferences in the expansion and smooth flow of value. The conditions necessary for uninterrupted long-term growth are impossible to meet, even in an economy conceptualised in very abstract terms as one of two sectors, one sector producing the means of production, the other producing consumer goods. As such, flows of value are unavoidably vulnerable to the threat of interruption and the capitalist social relations in which they are embedded are consequently endangered.[3]

2.3   Beyond the primary circuit of capital

The circuits of commodity, money and industrial capital encompass fixed capital investment in specific buildings and spaces (factories, offices, schools, banks, shopping centres and malls) and communications and transport infrastructure necessarily required for the production and realisation of surplus-value and the flow of value in and around the capitalist economy. In practice this can be a result of investment by state or private capital. State provision is motivated by market failure and the refusal of private capital to invest in specific items that are required as general conditions for accumulation to be possible, such as roads, because the rate of return is too low. Indeed, the successful reproduction of the primary circuit additionally requires the construction of houses for workers, schools, hospitals and other public facilities.[4] In this sense, the primary circuit of capital necessarily requires the material construction of a built environment and specific spaces in which the practices of the economy can be performed. However, the production of the built environment is also critically related to secondary and tertiary circuits of capital and to these as temporary solutions to problems posed by the 'over-accumulation of capital' (Harvey, 1989, 148). Harvey refers to the built environment as a product of the secondary circuit of capital while the tertiary circuit involves investments in education, health and welfare to improve the quality of labour-power and meets the legitimate demands of citizens for decent lifestyles and living environments.

The switching of capital between the primary and secondary circuits, and as a corollary between sectors and locations, mediated by private sector financial institutions and the public sector institutions of national states is critical to facilitating the circulation of capital. In phases of over-accumulation of capital in the primary circuit (characterised by high levels of stock, declining demand and falling rates of profit), capital will switch, speculatively, to the secondary circuit and to investment as fixed capital in the built environment. In this way, switching between circuits is linked to processes of speculative urban growth and more generally growth of the built environment in new housing areas, office blocks, commercial and shopping centres and so on, often specifically targeted as new spaces of consumption and sale. As such, switching between circuits of capital is intimately linked to the creation of new sorts of economic space and decisions about new public sector investments in the built environment. These can in turn have a decisive influence upon private sector locational decisions by virtue of the way that they affect possibilities for profitable activities. This process is, however, contradictory as a way of maintaining the flow of capital. New capital investment resulting from switching between circuits produces new forms of built environment and fixes capital in those spaces for as long as is required to amortise it.

2.4 From values to prices and the disciplining role of money

Value analysis is designed to reveal the defining relationships of a capitalist economy, not to describe social reality as experienced by people living in particular capitalist societies. The routine performance of the social relationships of production, exchange and consumption and the day-to-day conduct and market transactions of a capitalist economy (such as declaring profits or paying wages) are conducted in prices, not values. Economic agents freely enter into market relations mediated by monetary prices. Money thus serves as both a medium of exchange and a measure of value, though one that does not equate to values defined in terms of socially necessary labour time. In fact, it *never* is nor can it be that money prices are perfectly correlated with values defined via socially necessary labour time. For while money is a representation of socially necessary labour time, price the money name of value, money is always a slippery and unreliable representation of value (Harvey, 1996, 152). The discrepancies between supply and demand in markets result in commodities being exchanged at prices that diverge from their values. As production conditions diverge from the social and technical averages, the amounts of labour time embodied in commodities deviate from the socially necessary amount that defines the value of a commodity. Commodities thus contain varying amounts of labour time but are sold at the same market price while money prices typically diverge from exchange values. As a result, there is a redistributive flow of value between sectors and companies via processes of competition. This is also important in relation to the systemic dynamism of capitalist economies and their historical geographies of production and uneven development, and to processes of 'creative destruction' as firms seek competitive advantage via innovation and revolutionising the what and how of production.[5]

There has been considerable debate as to the 'correct' way to connect price and value analyses. For some, the critical issue is the quantitative transformation of values to prices, reflected in the history of the 'transformation problem', and more generally the issue of the validity of value analysis (Rankin, 1987; Roberts, 1987). An alternative approach is to recognise that these are concepts of qualitatively different theoretical status. Values and prices are indicative of the way in which capitalist social relationships unite qualitatively different types of labour in the totality of the production process. Massey (1995, 307) trenchantly argues that the law of value is useful for thinking through the broad structures of the economy and for forming the 'absolutely essential basis for some central concepts – exploitation for instance'. As such, value theory describes a specific set of social relationships in which exploitation is a process of extracting surplus labour that can only be understood in the context of the wider social forms constitutive of capitalism as a system of commodity production. Value theory therefore helps elucidate social relationships specific to capitalism. However, attempts to use it as a basis for empirical economic calculation are misconceived and doomed to failure. Indeed, 'the byzantine entanglements into which the "law of value" has fallen make it … unusable in any empirical economic calculus' (Massey, 1995, 307). It is therefore important not to confuse values and prices conceptually or seek to equate empirical data measured in prices with theoretical constructs defined in terms of values. The significance of value analysis is that it focuses attention upon class relationships, the social structures that they help to define and the resultant flows of value through which the reproduction of the economy is secured.[6]

While debate continues about the relationships between value and money, and the appropriate uses of value and monetary concepts and measures in analysis, the transactions of capitalist economies continue to be conducted in prices and monies. According to Dodd (1994), all monetary systems share five essential abstract qualities. First, they encompass a system of accounting to enable money to function as a medium of exchange, as a store of value and as a measure of account. Secondly, they require a system of regulation, to defend and protect the integrity of these three functions. Thirdly, they need reflexivity – that is, confidence that deferred payments will be met, based on expectations from past behaviour. Clearly, this places a premium on trust and the predictability of behaviour and outcome. Fourthly, they are based upon sociality, enabling information about money and value to be exchanged between people and organisations. Finally, they are based on spatiality – that is, recognition that monetary networks have specific types of territory, that monies are territorial and relate to particular political spaces of governance. While the first of these qualities can be thought of as representing the formal quantitative functions of money, the next four refer to the socially variable ways in which these three functions can be guaranteed and in which flows of capital, money and value around the economy can be assured. As such, it is reasonable to expect that they be guaranteed in particular ways within capitalist economies, and that there will be variations between different models of capitalism in this regard, as well as temporal changes as socio-technologies alter. Consequently, one might expect regulatory systems and concepts of trust to vary over time/space, along with modes of sociality as transport technologies and ICTs become more sophisticated. For example, Du Gay and Pryke (2002, 4) emphasise the importance of trust in contemporary, sophisticated monetary economies. Referring to expert knowledges relating to complex innovative financial products in major financial centres, they suggest that 'the pricing models and financial engineering that such experts compose and put in play ... display a collective trust in a monetized future that allow the so-called new financial instruments such as derivatives to work as forms of money'. In short, the creation of new forms of money and monetised commodities necessarily depends upon trust in, and the predictability of, a monetised future.

The money form pre-dates capitalism by several centuries (Davies, 1994) but as capitalism emerged and evolved, local and regional currencies were eliminated as national monies became established, a key marker of the competencies and authorities of national states and a necessary condition for the creation of national markets. This nationalisation of economic practices created resistances and barriers to the operation of the law of value, as well as establishing conditions for flows of money and value between national monetary systems. From the early 1970s, however, with the erosion of the Bretton Woods system of fixed exchange rates between national currencies, spaces of monetary and financial regulation increasingly approximated to global spaces, de-nationalising monies and weakening the barriers that they posed to financial flows and flows of money capital. Such global flows are controlled from a small number of global cities and off-shore centres (Leyshon, 2000). The emergence of floating exchange rates and digital money in the last quarter of the twentieth century progressively further undermined the capacity of national governments and central banks to control their own currencies, thereby facilitating the untrammelled workings of the law of value.

Increasing digitalisation and the expansion of electronic currency trading systems provided technological preconditions for the emergence of more global currencies and globalised financial markets but also greatly enhanced the instability of currency trading systems. Jessop (2000, 4) elaborates on the growing globalisation of the economy – 'the crisis of Fordism'[7] – relating shifts in monetary relations to flows of value and to movements between and within circuits of capital. Flows between circuits may be related to moments of crisis and the interruption of the smooth flow of capital and value around its varied circuits. Jessop argues that a major contributory factor to the emergence of this particular 'crisis of Fordism' was the undermining of the national economy as an object of state management, notably through the internationalisation of trade, investment and finance. This led to a shift in the primary aspects of its two main contradictions.

Thus the wage (both individual and social) became increasingly regarded as an international cost of production rather than a source of domestic demand; and money came increasingly to circulate as an international currency rather than as national monies, thereby weakening Keynesian economic demand management on a national level. This shift in the primary aspect of the contradiction of the money form is related to the tendency for the dynamic of industrial capital to be subordinated to the hypermobile logic of financial capital and the tendency for returns on money capital to exceed those on productive capital. This latter tendency may lead to increasing flows of money capital into the secondary circuit and the built environment and into other speculative forms that offer the promise of higher rates of return. As such, developments in the realms of money have impacts upon value flows in the economy and the distribution of surplus-value within and between circuits, which in turn has implications for crises and developmental trajectories. Higher returns to money capital necessarily reduce returns to productive, industrial capital – and the moment of production in that circuit is the origin of surplus-value. Consequently, withdrawal of capital from the productive circuit affects the reproduction of the entire process of value production and value flows.

Because the contemporary world is dominated by capitalist relations of production and such relationships are expressed in terms of monetary relations and prices, the conventional money form and monetary relations constitute a Foucauldian disciplining technology which can powerfully influence developmental trajectories and geographies of economies. The money form is significant in that it is a means of carrying and imposing the asymmetric power relations of capitalist social reproduction, which are extended and stretched over space and imposed upon people, not least through the associated practices of regulation, including monetary regulation, of capitalist states.

2.5   Capitalising knowledge, flows of knowledge and the 'weightless economy'?

A necessary condition for knowledge to be capitalised and commodified is transformation from tacit to codified knowledge that can flow freely between people via a variety of media, but most significantly in digital form. Two further and related conditions are necessary for specialised markets for selling knowledge-embodied products to emerge (Athreye, 1998, 13–28). The first is technological convergence, which in turn requires the emergence of new generic technologies. This allows technological knowledge to be freed from its particular context and sold in specialised ways. Secondly, there must be recurring and reasonably frequent transactions. The emergence of markets in technological knowledge is, therefore, spatio-temporally specific, confined to periods of technological convergence.[8] However, in enabling knowledge to become commodified in this way, Intellectual Property Rights (IPR) legislation also imposes serious constraints upon productive research, flows of knowledge and the free exchange of ideas and, as such, may stifle the development of innovation and the forces of production (Bowring, 2003, 116).

Alongside the growing (contested) claims about globalisation of the economy, there have been parallel claims as to the growing importance of knowledge and information flows in an (allegedly) weightless, de-materialised 'new' economy, particularly in terms of the extent to which knowledges can be digitised, commodified and capitalised. There are strong claims to the effect that this new economy operates in a complex, non-propinquitous, multidimensional cyberspace, with novel spatial dynamics grounded in the possibilities that cyberspace offers for simultaneous co-location of myriad entities and relationships (Leinbach and Brunn, 2001). There undoubtedly has also been growth in the importance of some sorts of knowledge and information in relation to commodity production and of the production of a range of 'symbolic' commodities. Switching capital into the third circuit is particularly important in this context in so far as claims about the centrality of knowledgeable workers and flows of knowledge are valid. Even so, claims as to the increased importance of flows of knowledge and information for economic performance have only limited validity, sectorally and spatially.

The selectively increased importance of flows of knowledge and information in some sectors of capitalist economies has highlighted the importance of processes of knowledge creation and flows of information within firms via a range of types and ways of learning. Learning and innovation involve complex circuits of knowledge and information between as well as within firms. The growing distanciation of many economic relations within an increasingly spaced out economy, as the locations of activities both within and between firms become further separated by physical distance, is made possible by increasing digitalisation and other improvements in ICT and transport technologies. Flows of information both increase in volume and in distance travelled, as do flows of people as sites of embedded and tacit knowledges. More generally, there is evidence of the creation of new global circuits of intellectual capital (Thrift, 1998).

Recognising these recent changes, it is nevertheless important to acknowledge that knowledge has not suddenly become economically important. The economy has always depended on knowledgeable workers and flows of knowledges and information. Indeed, an 'un-knowledgeable economy' and economic practices and performances that are not based on knowledge are literally inconceivable. What is at issue is the changing significance of knowledge, the new ways in which knowledge is economically important, the varying 'mixes' and types of knowledge, and the routes through which they flow into the production of *any* commodity. For example, Allen (2002, 39–40, emphasis in original) emphasises 'the symbolic basis of *all* forms of economic knowledge'. Furthermore, 'different economic activities play across a variety of symbolic registers – abstract, expressive, affective and aesthetic – and *combine* them in ways that render sectors distinctive'. Symbolic knowledge is not, therefore, confined to the production of cultural commodities and may have become relatively more important across a range of other commodities. Conversely, producing symbolic outputs typically requires substantial material underpinning and infrastructure, not least in creating specific settings to enable co-presence of producers and consumers. For example, IT services require particular sorts of building, computer, network connection and electricity, which in turn requires some form of fossil fuel-generating technology. Complex connections between different bits of commodity production allow the production of new 'symbolic commodities' rather than de-materialised commodities emerging in a digitalised, weightless economy. The material basis and weight remain critical, albeit distanciated from the particular sites from which flows of information and knowledge emanate.

Recognising that all economic practice is based on knowledge, Jessop (2000, 2) suggests that 'what is novel in the current period [of capitalist development] is the growing application of knowledge in developing the forces of production and the increased importance of knowledge as a fictitious commodity in shaping the social relations of production'. For example, one indication of this is the expanding volumes of patents awarded to companies involved in biotechnology and bio-engineering, which are positioned at the forefront of the new 'knowledge economy' in which 'information and ideas have become critically important economic assets' (Bowring, 2003, 118). At least three processes are involved in transforming knowledge into a fictitious commodity. First, the formal transformation of knowledge from a collective resource ('intellectual commons') into intellectual private property as a basis for revenue generation (for example, as a patent). For instance, companies have sought to transform the knowledges that indigenous peoples have of plants and animals into patented and privately owned knowledge to form the basis of commodity production in agriculture and cognate activities. Secondly, the formal subsumption of knowledge production under exploitative class relations through the separation of intellectual and manual labour and the transformation of the former into alienated wage labour, producing knowledge as an exchange value rather than as a use value. Thirdly, the real subsumption of intellectual labour and its products under capitalist control through their commodification and integration into a networked, digitised production-consumption process controlled by capital, of information produced by a firm not for its own use (as a use value) but to sell to another to deploy in its production process (as an exchange value).

Thus the distinctive features of recent developments in circuits of knowledge and intellectual capital relate to their global reach and speed of flow within them, changes enabled by innovations in ICT and the deployment of different combinations of knowledge in commodity production. It is these changes rather than knowledge and learning *per se* becoming distinguishing features of the capitalist economy that are crucial. There are, however, limits to such processes. Cyberspace is not a 'neutral third space' between capital and labour, market and state, public and private. Rather, it is a new terrain on which conflict between these forces, institutions and domains can be fought out. 'An oft-cited expression of this contradiction' is the institutional separation of hypermobile financial capital, circulating in an abstract space of flows, from industrial capital, still necessarily territorialised and fixed in space. But it also appears in the individual circuits of financial, industrial and commercial capital, as well as within their interconnections. For however much capital migrates into cyberspace, like all capital 'it still depends on territorialisation' – that is, on materialisation in specific spaces, whether they be global cities or industrial districts. Indeed, 'even e-commerce needs such an infrastructure, even if it involves a "celestial jukebox" sending digitised music on demand' (Jessop, 2000, 4).

2.6   Alternative definitions and flows of value

While national monies help create conceptual and practical spaces for different forms of capitalism, the possibilities of informal local currencies create space for alternative definitions of 'the economy' and of circuits and flows of value within the interstices of capitalist economies left uninhabited by capitalist social relations. Local currency systems differ from the state-regulated monies of the formal economy on each of the abstract characteristics of monetary systems identified by Dodd (and described above). They constitute different circuits of value, much more localised and spatially constrained – by design and intent – than those of the mainstream economy and its monies and deliberately challenge its concepts of value and processes of valuation.

Alternative local currency systems and concepts of value are designed to enable socially desirable processes of production, exchange and consumption to be conducted. For example, Time Dollars, potentially at least, pose a radical challenge to the value concepts of the mainstream economy as they equalise differences of skills and remuneration through the measure of time. They seek to challenge and overcome the power relations and resultant constraints inherent in mainstream monies and their Foucauldian disciplining effects. In the absence of alternative currency systems, therefore, such challenges would not be mounted. This is for one of two reasons. First, they would be impossible, as they would be incompatible with the financial and social relations of the mainstream formal economy. Secondly, they would be transformed in morally or socially unacceptable ways if conducted within the parameters of the mainstream formal economy. These alternative economic activities only become possible because of agreement by participants in the local currency system to engage in mutual exchange. Such socially constructed currency systems seek to encourage and promote economic practices that conform to norms that are morally and socially acceptable to those who participate in and administer them and that are seen to bring about progressive economic and social change. There are, however, important differences between local currency systems in this regard. In some, moral obligation is minimal or wholly absent, whereas in others the alternative moralities of exchange are pre-eminent. The cultivation of personal trust and creating alternatives to the impersonality of the formal economy via the promotion of communitarianism, mutuality and self-help, is a critical motive for participation in the latter schemes. Exchange within the local currency system involves flows of values defined in ways that differ, often radically, from the mainstream. They both embody and permit alternative and multiple decommodified circuits of production, exchange and consumption.

However, because at least some local currency systems represent alternatives to formal monies and their underlying concepts of value and value relations, rather than as parallel systems to them, they can be seen to pose a radical challenge to the mainstream. While they offer scope for the practice of alternative (concepts of) values, and so open up a range of political possibilities, there is an uneasy, hotly-contested fault line between such local currency systems and formal circuits of social reproduction. As such, they are likely to be confined to spaces beyond or on the margins of the mainstream, or in interstices within the mainstream that capital has no desire to flow into and through, rather than constituting an alternative mainstream. Not least, they are typically spatially circumscribed and, in that sense, local currency systems. As such, these alternative flows of value occur alongside one another and also alongside – albeit in an uneasy relationship to – flows of values and circuits of capital in the formal mainstream economy. Furthermore, the extreme institutional thickness of many local currency systems may pose a substantial barrier to their developmental potential and to their becoming more than institutions of social solidarity and emerging as significant circuits of social reproduction based around flows of alternative values.

2.7   Summary and conclusions

In this chapter I have discussed the ways in which capitalist economies can be thought of in terms of circuits of capital and flows of value, recognising that conceptions of value are contested and depend both upon the specifics of the social relations of a given economy and upon the ways in which these are theorised. While capitalist conceptions of value are dominant within capitalism, they co-exist alongside other views of value and the practices through which they are (re)produced. Money has a particularly important role in lubricating circuits of capital, but also forms a critical disciplining technology that strongly influences these circuits and the movement of capital within and between them. Marxian political economy seeks to penetrate the appearances of market exchange and values defined as market prices and uncover the origins of profit in the commodification of peoples' capacity to work as labour-power. Capital ceaselessly searches to draw people and things into its concepts of valuation and flows of capital, seeking (literally) to capitalise and commodify elements of both nature and human beings, issues that are explored further in the following three chapters.

Notes

1   Thus the circuit of productive capital involves fixing capital in material spaces of production, via private or public sector investment. Investment in general conditions of production is typically undertaken by national states (see Chapters 6 and 7).

2   The appropriation of raw materials involves a different type of labour process from that of transformative industrial production (see Chapters 4 and 7).

3   This knife-edge movement along a crisis-prone trajectory directs attention to three things. First, competition between companies as the motor of industrial dynamism. Secondly, the activities of the state in making production possible. Thirdly, the varied ways in which this is done. The first two are discussed at length in R. Hudson, 2001, Chapters 3, 5 and 6. The third is a recurrent theme in what follows.

4   Such facilities may be provided as commodities by the private sector or in (partially) de-commodified form by the state (see Chapter 6).

5   See R. Hudson, 2001; Chapters 5 and 6.

6   Consequently, many aspects of use values cannot be captured in value categories.

7   'Fordism' encapsulates a particular set of regulatory arrangements that privileges the national space and scale (see Chapter 6).

8   Consequently, in many firms and industries production remains grounded in specific technologies and tacit knowledges (see Chapter 7).

3 Flows of Materials, Transformations of Nature

3.1 Introduction

The economy can be conceptualised as a series of material transformations and flows (biological, chemical and physical). These encompass extracting raw materials from nature and converting them to socially useful resources, converting living plants and animals into inputs to production, selectively shaping these life forms via a variety of technologies, and consuming and eventually discarding the commodities produced. Such material transformations are marked by feedback loops, symptomatic of the complexity of interactions between people, nature and things and their diverse effects (both intended and unintended), and the emergent properties that characterise such complex systems. While the economy can be considered in terms of biological/chemical/physical transformations *per se*, these are shaped in specific ways by social relationships, and so vary within and between capitalist and other social relations. Consequently, while capitalist production involves the production of commodities, it can never be simply the production of commodities by means of commodities since at some point production necessarily involves appropriation from nature and the grounding of the economy in nature.

3.2 The economy as processes of material transformations and natural limits to economic life

3.2.1 Some basic concepts of material transformations

Economic activity involves the application of human labour, deploying a variety of artefacts and tools, to transform and transport elements of nature to become socially useful products. However, these processes unavoidably give rise to unwanted by-products and wastes as inputs that do not emerge as desired products appear in these forms (Figure 3.1). Consequently, *any* form of production, transport and consumption has an environmental footprint (Jackson, 1995) and the economy can be conceptualised as flows of energy and chemical and physical transformations of elements of nature. The laws of thermodynamics provide key

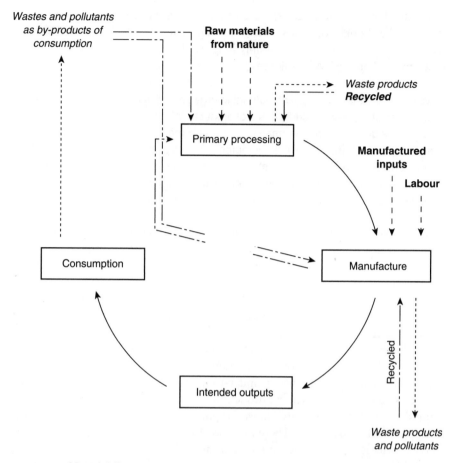

FIGURE 3.1 Material flows

insights in understanding these processes of material transformations. Crucially, thermodynamics characterises *any* material transformation as dissipative of energy and conservative of materials. These laws impose limits upon any form of economic activity.

Each industrial process and economic activity involves the *transformation* of materials and energy from one form to another. Thermodynamics provides very specific rules that govern these transformations; they cannot be altered or suspended by human intervention and in that sense set natural limits on social production and its relationships to nature.[1] The laws of thermodynamics state that energy is neither created nor destroyed during these transformations although it may change in physical form (for example, from kinetic energy to heat) and that the total mass of inputs to a transformation process is equal to the total mass of outputs. This identity holds at the level of individual atomic elements during (non-nuclear) material transformations. The economy can therefore be thought of as a sequence of steps, each of which 'is more or less a transient event, a temporary (possibly long-lived but temporary) use of some set of atoms and energy'. Moreover,

> we can postulate a universe of material/energy paths through the production, life, and dissolution of any product or set of products. We can also consider each path to be a sequence of transformations from one material/energy embodiment to another. We can view the whole of material industry as a network of such paths and transformations, connected at each end (extraction of materials and disposal of products) to the environment external to the process and product and at places in the middle (disposal of incidental waste). (Frosch, 1997, 159)

This, in principle, allows precise accounting of the environmental impacts of material transformations. It provides the conceptual basis for industrial ecology and the methodologies of life-cycle analysis and industrial metabolism[2] that seek to construct a set of accounts centred on the notion of mass balance.

In short, energy and matter are conserved during processes of transformation and there are methodologies that allow a precise tracing of both during these processes. By tracing the impacts of varying combinations of technologies of production and consumption, and of different levels and compositions of output, the ecological implications of economic choices about how and what to produce and consume can be better understood. Moreover, this could in principle be extended to consider where production occurs, for example in terms of companies' attempts to find 'spatial fixes' for pollutant and environmentally noxious and hazardous production. There are, therefore, considerable potential benefits in conceptualising economies in terms of industrial metabolism.

Although energy and matter are conserved, there are persistent fears about exhausting supplies of carbon fossil fuels and metallic minerals. The key to understanding this seeming paradox lies in the second law of thermodynamics, the basis of the economy of life at all levels (Georgescu-Roegen, 1971), and in the concept of entropy. This second law addresses the *loss of availability* of energy. As the same quantity of energy passes through successive transformations, it becomes progressively less available and so less 'useful' for production and human use as entropy increases,[3] assuming for the moment a simple closed eco-system.

However, complex economies are more appropriately conceptualised as open systems, with the tendency towards disorder and randomness countered by external inputs of energy. Energy is 'imported' to fuel economic processes, from the stock of carbon fossil fuels, and to a lesser degree from flow resources (such as solar energy, hydro-electric, wave or wind power). Low entropy fossil fuel reserves provide high-quality thermal energy through combustion. This can then be transformed into other forms and used, *inter alia*, to access resources that are unavailable to non-human species and deploy them in production and consumption of goods and services. But in developing in this way only a small fraction of the materials extracted from nature and mobilised in the economy are recovered and used.

Human societies and economies cannot escape the indeterminacies, uncertainties and constraints set by the laws of thermodynamics but 'it is quite another thing to treat them [these laws] as sufficient conditions for the understanding of human history' (Harvey, 1996, 140). However, these limits are non-trivial. Because the global ecological/economic system is complex and non-linear, its dynamic behaviour is potentially chaotic and its stability, its tendency to remain within its original domain, is indeterminate. Given the indeterminacy, there are good reasons to exercise the precautionary principle in considering relations between economy and environment. The global system is held in a steady state far from its thermodynamic equilibrium only by capturing and using radiant solar energy in various forms, notably as fossil fuels. However, it remains an open question as to whether *any* form of economy, *any* set of social relations of production, can develop effective regulatory mechanisms to contain the consequences of human intervention into the cycles of natural processes over the long term.

3.2.2 Beyond material transformations and the laws of thermodynamics: re-socialising economies and their relationships to nature

There are advantages to conceptualising economies in terms of material transformations but also limits to such approaches (Taylor, 1995). Most seriously, they abstract the economy from its socio-spatial context and the socially specific imperatives that generate particular socio-spatial configurations of economic organisation and practice. Material transformations therefore offer at best a one-dimensional, partial and restricted view of the links between economy and environment and the dynamics of human-induced environmental change.

Material transformations are mediated in particular ways as a consequence of dominant social relations. In capitalist economies there is a strong imperative to conquer and dominate nature to further accumulation. Capital in many ways still regards the natural environment as an unproblematic source of raw materials, of renewable flow and non-renewable stock resources that can be drawn upon and, in many instances, literally capitalised. Conquering nature is seen as one over-arching route to producing profitably and emancipating society from natural constraints. There have been massive demands upon the natural resource base, especially the finite stock of non-renewable resources such as metallic minerals but especially carbon fossil fuels that are only reproducible over geological time. This prospect of resource exhaustion raises a related and familiar, but nonetheless quite critical, question: what converts natural materials into natural resources? Harvey (1996, 147) offers a 'relational definition' of a natural resource as a 'cultural, technological and economic appraisal of elements and processes in nature that can be applied to fulfil social objectives or goals through material practices'. Thus natural resources are not naturally resources. Natural materials become or cease to become resources under specific combinations of social and technical conditions. At least three conditions must be met simultaneously for natural materials to 'become' resources. First, the existence of a need, or more specifically in a capitalist economy, effective demand that will generate profits from production. Secondly, the existence of appropriate enabling technology and relevant knowledge – both know where and know how. However, there is an interdependence between technology and the degree of concentration of resources – as measured by ecology or geology – as this strongly influences the quantity of labour and non-human energy needed for transformation from raw material in nature to natural resource (Deléage, 1994, 39). Thirdly, there must be political control and the guarantee of property rights to exploit resources. If any of these conditions cease to hold, resources may 'unbecome', reverting to the status of 'neutral stuff'. Elements of animate and inanimate nature once again become simply naturally occurring materials or conditions, often with severe localised socio-economic impacts (Beynon et al., 1991).

Social relationships other than those of capital also mediate linkages between people and nature. These exist or have existed in parallel to the social relations of capital, either as alternatives to, or within the interstices of, them. For example, state socialist societies prioritised the subjugation of nature through the activities of the state, allegedly in pursuit of socialism and collective well-being. However, events such as Chernobyl (Marples, 1987) and the ecological tragedy of Lake Baikal revealed that state socialism often had a more deleterious environmental footprint than did capitalism. This raised serious questions, albeit belatedly, about appropriate socialist ecological perspectives (Harvey, 1996, 146). Furthermore, there is a variety of non-capitalist 'indigenous societies', such as the Australian aborigines living as part of 'nature', seeking to reproduce nature 'as it is' (Massey and Catalano, 1978). Furthermore, within late modern societies there are views that echo those of such 'indigenous' peoples, expressed in various shades of 'green' views on people/nature relations, as people seek to live with and/or conserve 'nature' and develop balances between the natural and the social that are reflected in the governance and regulation of material transformations. These views range from a faint blue green of technological fixes and market solutions to the dark deep green of eco-warriors who prioritise the preservation of an essential nature over the well-being of people.

Taking the long historical view, capitalism and state socialism have sought to displace indigenous societies as the dominant forces governing the form and processes of transformation. More recently, state socialism has collapsed, in part because of mounting opposition to environmental destruction from within such societies. This has left capitalism in an even more dominant position as the mode of economic organisation through which the transformation of nature is mediated, but one that has triggered oppositional movements that echo non-capitalist views as to the 'proper' relationship between people and nature.

3.3 Nurturing nature: cultivating plants, rearing animals and optimising conditions for their development, growth and commodification

Many inputs from nature to the economy are, in principle at least, renewable within human time scales, with much shorter periodicities of (re)production and many are the subjects and objects of a range of human activities that seek to modify nature and natural ecologies, such as agriculture, fishing and forestry. There is a great variety of pre- and non-capitalist forms of agricultural production, involving the domestication and cultivation of plants, and the taming of animals and animal husbandry. Speaking at a meeting of the Food and Agricultural Organisation in 1985, Lopez-Portillo (cited in Whatmore, 2002, 99) noted that:

Historically, all manner of civilizations have depended on, or created, hundreds of thousands of plant species and varieties for their daily sustenance, for health and hygiene, for clothing, shelter, for obtaining dyes and chemical, for symbols and progress. In sum, for the harmony and stability of their geography, their society, their culture.

In such circumstances, activities such as agriculture remained necessarily bound to the rhythms and times of natural biological and bio-chemical processes. As a result, production is subject to the vagaries of climate and nature as human labour can only seek to optimise natural growing conditions and the developmental transformation and growth of plants and animals within these parameters.

Consequently, capital has sought to shape these transformatory processes in specific ways. The penetration of capital into agricultural production and associated activities such as forestry and ranching has given a particular focus to the developmental transformation of plants and animals. The trajectories of transformatory processes of growth have been directed towards the commodification of plants and animals and the accumulation of capital. Capitalist social relations have penetrated many areas of plant and animal production, from the mundane (for example, the potato or tomato) to the more – to western tastes – 'exotic', both in terms of the cultivation of crops (such as mangoes and pineapples) and the rearing of animals, commodifying a wide range of plant and animal products. Often these developments are linked to changes in international trade agreements. For example, the tenth meeting of CITES (Convention on International Trade in Endangered Species) in 1997, in Harare, agreed to change the status of *Caiman litirosis*, so that Argentina could trade the skins of 12,500 ranched crocodiles in the period 1998–2000. This ranching process commences with appointed scientists collecting eggs from 'wild' crocodile nests. These are then incubated and hatched in commercial ranching facilities. Some 2,000 marked juveniles, aged 8–10 months, are returned to the 'wild' each year but the majority are retained, raised to kill size, and their skins (bearing the hallmark of CITES approval) are then sold in international leather markets to be transformed into 'exotic' fashion accessories (Whatmore, 2002, 29–30).

But such dependence upon natural biological processes remains subject to the vagaries of nature. This can and does lead to fluctuations in output and unpredictability in the processes of creating new strains and species and fluctuations in and unpredictability of profits. This is inimical to capital. At the same time, it is an opportunity for capital to penetrate and colonise these spaces and subsume these activities, damping down or eliminating variability and uncertainty via 'out-flanking' nature and producing more controlled and predictable conditions and ecologies of production. Capital has sought to outflank nature via two interrelated processes: appropriationism and substitutionism (Goodman et al., 1987). Appropriationism involves replacing previously 'natural' production processes by industrial activities. Substitutionism refers to the substitution of synthetic products for natural ones. In the specific context of food, these processes comprise a strategy progressively to eradicate biological and bio-chemical constraints on production. In the more general context of industrial production, they constitute an attempt to emancipate production from the constraints of nature via transforming natural environments to reduce or eliminate uncertainty and unpredictability. Processes of production, to varying extents depending upon the nature of the product, are unavoidably shaped by the biological, chemical and physical properties of raw material inputs and outputs and of the surrounding environment. As such, both the scale and predictability of production – and the profits to be made from it – are prey to the vagaries of nature.

However, attempts to reduce uncertainty and increase predictability systematically are limited in their effectiveness by the distinctive character of the labour process in capitalist agriculture. In agricultural labour processes, human labour is deployed to sustain or regulate environmental conditions under which plants and animals grow and develop. There *is* a transformative moment but transformation is brought about by naturally given mechanisms and processes. Human labour is applied primarily to optimise the *conditions for* transformations that are organic processes, relatively impervious to intentional human modification and in some cases absolutely non-manipulable. For example, the incidence of solar radiant energy is such a process and labour processes in agriculture are thus confined to optimising the efficiency of its 'capture' by photosynthesising plants or complementing it with artificial energy sources. Despite efforts to 'industrialise agriculture' production involves seeking to optimise natural conditions in relation to the growing requirements of particular species (Benton, 1989, 67–9).

Within agriculture, therefore, capital seeks 'localised' adaptive solutions to problems posed by the natural environment for predictable and profitable production. For example, capitalist development of food production has sought to outflank biological processes such as ripening and rotting via refrigeration and air transport. Other technological innovations allow production in a wide range of spaces through creating appropriate environmental conditions via techniques such as irrigation that make agriculture possible in deserts, hydroponics (replacing natural soils with a variety of growing media) to enable production in areas devoid of suitable natural soils, and the creation of the artificial environments of fish farms, glass and plastic houses, raising the temperature and regulating the humidity of local environments, sometimes with atmospheres enriched with carbon dioxide to enhance growth. Such intensification of agriculture substitutes energy for cultivable area, fossil fuels for solar energy. Often the creation of such environments is accompanied by the use of biological and/or chemical fertilis-ers. It may also involve the deployment of other techniques to stimulate growth, such as the use of mass-produced bumble bees to pollinate tomatoes grown in glass houses, seeking to adapt nature to the requirements of production (Harvey et al., 2003). Typically, pesticides are used to ensure that unauthorised non-human consumers do not eat crops before they can be sold. In these ways, as a result of innovations in communications, transport and production technologies and the ready availability of large masses of cheap labour-power in many peripheral spaces of the capitalist economy, links between natural ecologies and economic activities have been loosened and, often, obscured.

There are therefore strong systemic pressures to bring the diverse times needed by natural entities to survive, grow and reproduce more into line with the imperatives of capitalist production, resulting in often massive time/space dislocations for the former. Production of animals and plants becomes possible 'out of time' and often 'out of place' and agricultural production becomes increasingly globalised (Goodman and Watts, 1997). This enables 'exotic' trop-ical fruits and vegetables to appear on the shelves of supermarkets in affluent areas of Europe and North America throughout the year. Consequently, seasonality is increasingly rendered irrelevant as supermarket shelves become constantly filled with the same fresh products. Rather than attempt to dominate nature in some over-arching sense, localised solutions are devised that permit particular sorts of production in a range of time/spaces.

Such forms of localised environmental modification have increasingly been combined with computing and information technology control systems to allow more precise manipulation of growing conditions in artificially created growing spaces and of the times at which crops are harvested within them. In some cases this extends to allowing total traceability of individual items of fruit or vegeta-bles, to a level of detail that includes the employee harvesting a particular item and the precise time and location – the individual plant – at which (s)he does so. Such electronic control systems can also allow greater closure of energy flows within the production environment. Other technological changes seek more than just local adaptation, however.

3.4 From husbandry to bio-genetic engineering: capitalising nature and the real subsumption of nature by capital

As well as seeking to optimise local conditions for processes of bio-chemical and biological transformation and growth, people have sought to change the quali-ties of the plants grown and animals raised over longer periods of genealogical time. Historically, desired changes to the qualities of plants and animals destined for human consumption depended upon selective breeding in the streams of inter-generational and genealogical time. Hybrids, characterised by exceptional uniformity of height, width, fruit, yield and so on, are produced by cross-breeding distinct inbred lines until a match is found which yields progeny that exhibit unusual vigour ('heterosis', the tendency for the offspring of genetically diverse plants to perform much better than their parents). The mixing of plant genomes through sexual reproduction results in the loss of these superior agronomic traits in the next generation of plants, which suffer a subsequent drop in performance (Bowring, 2003, 117–18). As such, there is an unavoidable delay in assessing the success of outcomes, uncertainty as to outcomes as people seek to steer processes of natural selection in particular ways, and strict limits to the longevity of success.

However, the fact that hybrid plants must be renewed each year by farmers, especially those involved in large-scale mass production food systems in which the demands of mechanical harvesting, food processing and fickle consumers must be met, also opens up a potential annually renewable market and space of sale for capital to exploit, Consequently, companies moved into seed production, seeking to 'sterilise nature's own prodigious and normally renewable productive and reproductive power so as to prevent it from creating for those who work it their own means of production: seeds'. As such, Kloppenburg (1988, 93) notes that 'hybridisation thus uncouples seed as "seed" from seed as "grain" and thereby facilitates the transformation of seed from a use-value to an exchange-value'. As scientific and technological advances opened up further possibilities, companies sought to extend their control over the production of nature, seeking to create desired plants and animals via bio-engineering technologies of genetic modification and the creation of genetically modified organisms (GMOs) and transgenic species. Rather than seek to optimise 'local' growing conditions for existing species, genetic engineering seeks a more profound and global domination of nature by altering those species or by creating new ones.[4] Genetic modification has been widely used to alter the character of food products so as to improve desired characteristics (such as colour, size, shape, taste and longevity: Harvey et al., 2003). Genetically altered ragweed plants have been developed that clean soil contaminated by lead and other metals while micro-organisms have been developed to 'eat' toxic wastes generated in semi-conductor production (J. O'Connor, 1994, 157–8). More dramatically, genetic engineering can involve cloning existing species (most (in)famously, Dolly the sheep) or creating new transgenic species (for example, OncoMouse[TM]), further pushing back the constraints on the economy and human life and at the same time raising serious ethical and moral issues.

Bio-technological plant breeding 'recombines' the DNA of the target plant by altering its genetic sequence. In the case of transgenic plants it involves adding one or more genes from a donor organism. This recombinant (rDNA) process involves three key steps. First, isolation of the coding sequence for the genes associated with the desired trait, Secondly, replication and transfer of this gene (or genes) to plant cells. Thirdly, regeneration and developmental regulation of the gene in the target plant using conventional tissue culture techniques (Whatmore, 2002, 131). In the case of animals, similar processes are involved, inserting genes from one species into the embryos of others to create desired characteristics for specific purposes. For example, the transgenic OncoMouse™ was specifically created in 1988 by Du Pont and Harvard University, controversially engineered for medical research to carry and breed with human genes predisposing it to cancer. Genetic engineering effectively replaces genealogical time with instantaneous production in the space and time of the laboratory and its equipment in shaping evolutionary trajectories for plants and animals under the sway of capital, effecting an unprecedented and dramatic compression in the otherwise lengthy processes of biological breeding. Thus transgenic breeding departs from the familiar reproductive model, 'technologically assisting nature's own recombinant pathways by introducing new channels of genetic exchange, the human into the mouse, the fish into the strawberry, the protozoan into maize' (Franklin et al., 2000, 88). This is opening up the prospect of bio-engineering organisms with previously unthinkable combinations of genetic material and also of an unprecedented degree of control over the fertility, reproduction and development of living things (Bowring, 2003, 121).

In addition, these GM techniques also carry the promise – or threat – of further dramatic dislocations in the times and spaces of production of existing plants and animals. Using cloning and *in vitro* micropropagation techniques, companies will soon be able to mass produce, in carefully controlled growing environments in laboratories in the temperate climates of North America and Europe, high-value crops previously only produced in the tropics. Via this particular 'spatial fix' capital will be able to avoid the constraints and uncertainties of the unpredictability of the weather, seasonal variation, problems of labour, transportation and long-term storage of perishable goods (Bowring, 2003, 127).

Outside such laboratory environments, GM plants have already expanded rapidly, but with a very uneven geography. Since they were first commercially licensed in the USA in 1996, there has been an annual double digit increase in global acreage, so that by 2002 almost 59 million hectares were under GM cultivation, with over 70% of this accounted for by herbicide-resistant varieties. This was concentrated in Canada, the USA, Argentina and China, with, in all, cultivation in 16 countries.[5] Over 20% of the global crop area of soya, cotton, corn and rape was produced using GM varieties. Soya exemplifies the processes and conflicts involved in the rise of GM crop production and the relationships between capital and the creation of GMOs. GM soya has become one of a number of transgenic crops fabricated by Monsanto under the trademark Roundup Ready™, genetically modified to tolerate a broad spectrum (that is, indiscriminate) glyphosate herbicide Roundup®, Monsanto's flagship herbicide. By 1998, within two years of Roundup Ready™ being licensed for commercial planting, it accounted for almost 33% of all soybeans planted in the USA. The 'startling entrance' of GM soybeans 'signals the increasingly monopolistic impetus of

corporate efforts to enrol the seed into the service of other product lines'
(Whatmore, 2002, 130). Existing hybridisation techniques would have been
capable of breeding pest- and disease-resistant traits garnered from the diversity
of Asian soybeans into the 'branded hybrids' of industrial crops produced in
the USA. But genetic modification presented a quicker and commercially more
attractive vehicle for businesses that invested in hybrid seeds because of their
established interests in agri-chemicals (Kloppenburg, 1988). In Monsanto's case
the rationale for this was very clear: its patent on Roundup®, which accounted
for over 15% of its sales and over 50% of its profits in the 1990s, was due to
expire in 2000. Its fears were well-founded. In 2002 sales of Roundup® fell by
24% and Monsanto's total sales declined by 14%, resulting in a net loss for
the year of $1.7 billion. This provided a powerful imperative to consolidate its
lead in the production of genetically-engineered herbicide-tolerant seeds (Tokar,
1998, 257).

More generally, the opportunities offered by biotechnologies from the 1980s
triggered waves of acquisition and merger activity, redefining spaces of production, as companies sought to position themselves in this potentially very lucrative
market and shape the production of life in the interests of the production of
profits. By the beginning of the twenty-first century, four companies – Aventis,
Du Pont, Monsanto and Syngenta – accounted for virtually all production and
had a powerful oligopolistic control of the global market for transgenic seeds.
Moreover, some 77% of transgenic crops are modified to tolerate the herbicide
products of the companies that produce the seeds. This wave of mergers and
acquisitions and then joint ventures and strategic alliances between these and
other major firms is leading to the emergence of 'clusters of multinationals co-
operating in achieving complete command of the food chain', from the patent
protection of transgenic germplasm, through chemically assisted growing, to the
collection and distribution of harvests and their processing into food (Bowring,
2003, 109). This is indicative of the ways in which the development and growth
of plants and animals – and the scientific and technical knowledges on which they
depend – are increasingly shaped by the requirements of capital accumulation.

Monsanto exemplifies these processes of corporate reinvention and commodification of life, transforming itself from a chemicals company to a life
sciences company. In the 1980s Monsanto acquired several agricultural biotechnology and seed companies, becoming the second largest seed producer in the
world and the largest producer of genetically modified herbicide-tolerant
(GMHT) seeds.[6] However, this particular 'socio-material ordering' is only held
in place as a result of monopoly patents 'whose grip is re-inscribed by the sig-
nature that seals every purchase agreement each time a farmer buys Roundup
Ready™ seed' (Whatmore, 2002, 132). In similar fashion Monsanto inserted a
marker gene into its New Leaf Superior potatoes, reprogramming them to
produce their own insecticides. The marker, 'a kind of universal product code'
(Pollan, 1999, 11, cited in Franklin et al., 2000) allows Monsanto to identify its
plants and so enforce its patent licence to those who purchase its product to

grow potatoes to eat or sell, *but not to reproduce*' (Franklin et al., 2000, 73, emphasis in original). Furthermore, the cross-species transfer of DNA from human to mouse to create OncoMouse™ not only made the animal a transgenic life form but also a new form of private property. This insertion of DNA and marker genes means that as breeds have been partially de-naturalised through biotechnology, simultaneously '*brands have become renaturalisable in return –* for example, by being written into animals' genome' (Franklin et al., 2000, 91, emphasis in original).

This process may yet be taken further because of the development of 'terminator technology' (RAFI, 1999, cited in Bowring, 2003), a biotechnology to ensure genetic seed sterilisation. This involves creating transgenic plants that yield pollen or seeds made infertile by the release of a toxin, such as that expressed by a gene from the soil bacterium *Bacillus amyloliquefaciens*. When triggered by a promoter specific to a developmental stage of the plant, such as the drying out of the mature seed, this new gene expresses in the reproductive cells a toxin that makes it impossible for the proteins necessary for the matura-tion of viable gametes or embryos to synthesise. In effect, it introduces planned obsolescence into the plant. It thereby provides 'a technology ... a biological means of policing (or functioning in lieu of) patents on life forms' and to prevent seed saving from crops, especially soybean, rice and wheat, 'in which hybridi-sation has not been commercially viable' (Bowring, 2003, 136). Following the initial award of a patent jointly to Delta and Pine and the US Department of Agriculture in 1998, by 2000 all the major biotechnology companies had been issued with patents for terminator-type systems, although AstraZeneca and Monsanto responded to pressure by publicly stating that they would not (yet) commercialise such technology.

Corporate ability to brand and patent genetic material in these ways is a recent development, made possible by technological and regulatory innovations and the scope that they created for capital. Historically, a jurisprudential dis-tinction between 'physical' and 'intangible' property construed living things as belonging 'by their very nature' to the domain of the physical. This placed them beyond the compass of intellectual property rights (IPR) since they failed to meet the criterion of being a non-obvious and useful human invention (Hamilton, 1993, cited in Whatmore, 2002). It was not until the beginnings of an inter-national framework for Plant Breeders Rights (PBR) in 1961, with the creation of the International Union for the Protection of New Varieties of Plants, that some limited quasi-patent protection was offered to those breeding new plant varieties that are distinct and novel, stable in that they reproduce true to type, and uniform in that they are stable within a generation. This was reinforced in 1994 when the US Supreme Court amended the 1970 Plant Varieties Protection Act and ruled it illegal to sell saved seeds for planting purposes (Bowring, 2003, 120).

It was only in the 1980s, however, led by the US Supreme Court, that legislation and case law began further to shift these ontological co-ordinates by drawing new distinctions between the biological and micro-biological knowledge practices and objects that admitted bio-chemical in(ter)ventions and genetic entities into the company of things that can be patented (Correa, 1995). In 1987 the US Patent and Trademark Office ruled that all multi-cellular organisms, including animals, were eligible for patent protection. As such, a significant barrier to profitable production was removed, creating an attractive enlarged space of opportunities for capital. This enabled capital to seek out new sites of accumulation 'in the interior spaces of the bodies of women, plants and animals' (Shiva, 1997, 5).

These changes had further dramatic implications. Since IPR combines the universalising pretensions of science and law to effect a radical break with the past, they collapsed biological 'becomings into the here and now of invention such that a germplasm without a history is folded into a future of monopoly entitlement' (Whatmore, 2002, 109–10). Put slightly differently, the ways in which the reproduction of commodities, markets and capital now have their own explicit 'facts of life' is made particularly evident 'in the context of bio-commodities, such as genetically modified foods, where the brand is not only written into the product's DNA but is consumed in the double sense of being both purchased and eaten' (Franklin et al., 2000, 68).

This transition from selective breeding via seeking to shape processes of natural selection to more or less instantaneous bio-engineering of new transgenic GM varieties thus marks a decisive shift in the ways in which people seek to transform living things for their own particular ends and in the distribution of power to make such choices. As Franklin et al. (2000, 85–6) note, the heirloom variety seed 'indexes the most traditional uses of genealogy, mobilised to invite novel forms of personal consumption, self-health and political activism and environmental stewardship'. In sharp contrast, the patented clone and transgenic breeds manufactured by corporate agribusinesses and pharmaceutical companies 'signify the precise opposite to a wary public, both captivated and disturbed by their coming into being. In their making and their marketing, the new breeds depart significantly from conventional models of genealogy'.

Yet in their public relations material companies such as Monsanto seek to present this rDNA process as a straightforward extension of traditional breeding methods that simply allows for the transfer of genetic information in a more precise, controlled manner. This, however, ignores countervailing voices in a debate in the life sciences community that challenges this view and also ignores the 'trial-and-error' character of experimentation. One manifestation of this is the 'unintended effects' of GM soybeans. As Whatmore (2002, 134), not without irony, puts it: 'For all its precision engineering ... the GM incarnation of the soybean [does not] stay put in the germinal fabric of the seed or the field boundaries of the crop ... but is metabolised and redistributed through all manner of inter-corporeal relations in growing and eating practices'. In like manner, but more frighteningly, BSE emerged as an 'unintended consequence' of the intensive feeding regime of industrial cattle production. Thus the 'troubling spectres of fleshy mutability' that haunt the shadowy regimes between field and plate mass with particular intensity in the event of 'food scares'. Such events have become endemic to the relentless industrialisation of food production under the imperatives of capitalist relations of production over the last half-century and are emblematic of the threadbare fabric of trust (dis)connecting contemporary industrial food production and consumption (Griffiths and Wallace, 1998).[7]

This in turn led to feed-back effects and pressures to modify processes of transformation in agriculture, both in terms of recreating markets for non-GM seed varieties and of a concern to establish 'product traceability', especially in Europe. For example, Seeds of Change Inc, based in Santa Fe, has prospered, selling native seeds for home producers, 'heirloom variety' seeds that are ancient, organic and safe (Franklin et al., 2000, 85). Again, in May 1999 a European consortium of major food retailers formed to secure supplies of non-GM ingredients and derivatives. These countervailing commercial currents boosted the market for non-GM soya, primarily produced in Brazil[8] and Canada, raising the price and volume of sales of soya guaranteed *not* to be Roundup Ready™. In the process, 'this realignment of beans, contracts and devices' that could discriminate between and then keep apart GM and non-GM soya 'undermined the rubric of "equivalence" and dispelled the "impracticability" of their distinction' (Whatmore, 2002, 140). The 'traceability' of products is becoming increasingly important in relation to consumer perceptions of risk in the food chain and growing resistance to genetic modification among certain social strata and food retailers, as concerns over BSE and more generally genetically-modified foods graphically illustrate. This has lead to product innovation aimed at particular niche markets, especially in terms of organically produced organisms (Morgan and Murdoch, 2000). 'Product traceability' led to a concern to fix products to specific spaces of production, to brand product via specific space (and vice versa) as one way of securing traceability (though not necessarily quality), trust and regard. Major food retailers sought to appropriate local food production systems and spaces as a way of meeting consumers' concerns, minimising risk and avoiding liability, allied to marketing strategies that emphasise that their products contain no genetically-modified material.

In short, life science companies have invested in genetic engineering and the development of GMOs as a way of creating continuing monopoly markets for other products that they produce. These companies sought to claim full and internationally recognised patent protection for humanly 'invented' life forms and, as a final guarantee, the creation of organisms with built-in planned obsolescence – that is, the inability to reproduce. The biotechnology companies then sought to represent this as a simple extension of established processes of biological evolution rather than as a sharp qualitative break with them as the production of life is engineered under the sway of capital rather than evolved within the genealogical time of natural processes. However, these processes of genetic modification may be problematic, with impacts that are both unintended and unwanted, revealing emergent effects that in turn raised doubts and fears about transforming living matter in this way. Thus 'the newer biological technologies have been "sold" within a voluntaristic-Promethean discourse which has inevitably occluded or rendered marginal the limits, constraints and unintended consequences of their deployment in agricultural systems' (Benton, 1989, 68). As a result, they have contributed to a corrosion of public trust in scientific opinion and expertise. In turn, however, this has created commercial (and other) pressures to return to more established and 'natural' forms of agricultural production and evolutionary procedures for the development of life forms.

3.5 Capitalist relations of production and the production of nature: from first to second nature

Since the emergence of industrial capitalism there have been spectacular developments in science and technology and their application to the practices of the economy. Science and technology have not simply been systematically applied in production, but their development has been increasingly and explicitly focused on production for profit while the production of such knowledge becomes in part commodified. The transformation of nature has assumed qualitatively new dimensions. Relations between people and nature have been progressively mediated and shaped via socio-economic and socio-ecological institutions specific to capitalist production. The character of capitalist class relations defines a specific form of relationship with the natural world. The abstract logic that attaches to the creation of value and capital accumulation structures the form of relations with nature. Abstract determinations at the level of value are continually translated into concrete social activity involving interactions between people and nature (Burkett, 1997). This produces a very complex determination of relationships between people and nature.[9] Nature and society are indissolubly linked in specific ways as capital circulation and ecological processes intertwine to create complex environmental transformations (Harvey, 1996, 59). The drive for profits and, moreover, for increasing profits in successive accounting periods, shapes the appropriation of nature. As a result, there is a strong tendency to remove entities from their eco-systemic contexts. No part of the earth's natural environment is immune from such dislocation and transformation. There are certainly technological limits to the extent to which the effects of capitalist social relations can in practice penetrate beneath, say, the earth's surface in search of minerals at any given point in time. Such limits are, however, constantly being pushed back.

As capitalist relations have penetrated increasingly deeply, widely and spatially, they have decisively shaped connections between nature and society, both directly and indirectly. While historically pre-dating the emergence of capitalist production, the distinction between 'first' and 'second' natures was central to the development of production for exchange and, in turn, was increasingly eroded by it. Capitalist production increasingly produces nature 'from within', continuously redefining relationships between 'first nature' and 'second nature', expanding the scope of the latter at the expense of the former. In part this involves 'capitalising nature', designating as valuable stocks of erstwhile 'uncapitalised' aspects of nature, enabling capital to delineate clear property rights over natural domains and so facilitate their 'highest and best' use, as defined by the logic of capital (M. O'Connor, 1994b, 144). For example, patenting plants and seeds that previously were part of the commons of indigenous societies transforms them to private property with associated economic rights. More generally with the transition from first to second nature 'we enter a world in which capital does not merely appropriate nature then turn it into commodities … but rather a world in which *capital remakes nature* and its products biologically and physically (and politically and ideologically) in its own image. *A pre-capitalist nature is transformed into a specifically capitalist nature*' (J. O'Connor, 1994, 158, emphases added). Moreover, the residual first nature is increasingly humanised, even if its components remain 'wild', as their use and management become subject to detailed human control – for example, in rivers, forests, grouse moors or 'big game' parks.

As first nature is increasingly produced from within and as a part of second nature, these natures are themselves redefined. With production for exchange, the difference between them becomes simply the difference between non-human and humanly created worlds. Once people produce first nature, however, this distinction ceases to have substantive meaning. The significant distinction now becomes that between a concrete and material first nature and an abstract second nature, derivative of the abstraction from use value that is inherent in exchange value. The same piece of matter thus exists simultaneously in both natures. As a physical entity, it exists in first nature and is subject to the laws of biology, chemistry and physics. As a commodity, it exists in second nature, subject to the law of value and market movements. Material nature is thus produced via socially organised human labour, subject to the determination of the imperatives of second nature, the incessant drive for profits that define capitalist relations of production. Thus 'human labour produces first nature, human relations produces the second' (Smith, 1984, 55). More specifically, the social relations of capital produce second nature as natural, cultural and social impediments to the circulation of capital are removed. This 'usurpation' of space is simultaneously 'the production of space' and the construction of a 'second nature' (Altvater, 1994, 77).

This social production of nature has important implications for the treatment of nature in the economy and the process of capital accumulation. Specifically,

> through the capitalization of nature, the *modus operandi* of capital as an abstract system undergoes a logical mutation. What formerly was treated as an external and exploitable domain is now re-defined as a stock of capital. Correspondingly, the primary dynamic of capitalism changes from accumulation and growth feeding on an external domain, to ostensible self-management and conservation of the system of capitalised nature closed back on itself. (M. O'Connor, 1994b, 126)

Indeed, M. O'Connor (1994b, 144, emphasis in original) further claims that the *modus operandi* of modern capital in its 'ecological phase' is not profit as such but 'semiotic domination. What matters is to *institute socially* the commodity form', thus representing all nature (including human nature) as capital, *ipso facto* in the service of capitalism as a legitimate social form. Looked at systemically, the pricing of a good or the successful capitalisation of an element of nature signals a 'semiotic conquest', namely 'the insertion of the elements and effects in question within the *dominant representation* of the overall capitalist system activity'. This has 'an undoubted "use value" for the project of the reproduction of capital as a *form of social relations*'.

Consequently, the distinction between first and second natures is increasingly rendered obsolete by the development of capital. As such, second nature increasingly encompasses the material world of fixed capital (the built environment, forces of production and so on), the social world of institutional formations that make production possible, and the discourses propagated about both. Because of the expansion of capitalist production, more elements of nature, previously unaltered by human activity, have been transformed to become elements of a socially-produced second nature. The production of first nature from within capitalist social relations and as part of second nature results in the production of nature *per se*, rather than of first or second nature in themselves, becoming the dominant reality. Thus the development of capitalist forces of production, to a degree, emancipates people from the domination of nature; however, this emancipation is integrally linked to antagonistic social relationships of production, grounded in the subordination of labour to capital.

The immediate goal is not the production of nature but the production of profits. Much of the production of nature is therefore an unintended and uncontrolled by-product of processes of capitalist production. Nevertheless, the production of nature in this way clearly requires the development of particular sorts of 'scientific' knowledge, which provide a cognitive basis for appropriating and transforming nature into socially useful products. People construct natural laws based on scientific discovery and investigation that can subsequently be applied in production.[10] By implication, the production of such laws also means recognition of their limits and so of natural limits upon the economy, irrespective of the particular forms of social relationships within which it is organised.

3.6 Summary and conclusions

In this chapter I have considered various aspects of the relationships between economy and natural ecologies, of the grounding of the former in the latter, and of the implications of this for shaping processes of transformation of elements of the natural world. These vary both in form and time period, from the chemical and physical transformations of finite stocks of mineral 'natural resources' to the biological transformation of life forms in a range of ways from the genealogical time of selective breeding to the instantaneous laboratory time of genetic engineering. This is the latest expression of a well-established tendency for capital to seek to move from appropriating an external first nature into the orbit of circuits of capital to producing nature as second nature within the social relations and circuits of capital. It can be seen as part of a long-term tendency to seek to 'master nature' in pursuit of profit. There is, however, a permanent tension between the social imperatives of capitalist production and the grounding of economy in nature, which renders it unable to escape the limitations of the laws of thermodynamics and increasingly susceptible to the complex emergent feedback effects of biological and bio-chemical transformation on life forms. Relationships between economic activities and practices and nature are prone to generate unintended consequences, reflecting the complexity of interrelationships between natural and social systems.

In summary, people, their societies and artefacts continue to be subject to the limits imposed by 'natural' laws and processes. No matter how efficient (in terms of energy and materials transformations) the organisation of economic processes and how far human society seemingly is emancipated from the constraints of nature, these limits remain. They are an unavoidable aspect of the human condition, irrespective of which particular social relationships of production happen to be dominant in a particular time and space. It is important to recall that a defining feature of capitalist relations of production is that they tend to undercut the conditions that make production possible, both in the worlds of first and second nature. As such, capitalist production threatens its own future viability via its rapacious appetite for natural resources, its incessant pressures to treat the natural environment as a free waste dump for pollutants, and its interferences in evolutionary processes as it seeks to erode genealogical time and genetically modify organisms. As Smith (1984, 62, emphasis in original) stresses, '[t]he production of nature should not be confused with *control* over nature'. Despite advances in scientific and technological knowledge, significant elements of first nature remain beyond human influence and control and pose risks to people. For example, earthquakes and hurricanes continue to wreak havoc on human societies, with no foreseeable prospect of their becoming internalised as part of second nature. Equally, prediction should not be confused with control over nature, even when it can be achieved. While predictive power is one criterion against which the theoretical adequacy of natural laws is judged, successful prediction depends upon a series of side conditions being satisfied. They may not be, but even if they are, prediction does not necessarily translate into control.

Notes

1 Benton (1989, 58) emphasises that 'in the face of realities which are genuinely invulnerable to human intentionality, adaptation by modifying or even abandoning our initial aspirations [to control nature] is to be recognised as a form of emancipation'.

2   Industrial ecology is based upon a normative claim that complex industrial systems ought to mimic natural systems (Scharb, 2001, 3). Industrial metabolism involves constructing a balance sheet of physical and chemical inputs to and outputs from a specific facility, industry or sector. A 'life-cycle analysis' is a cradle-to-grave mapping of the material inputs and outputs associated with producing goods and services.

3   Entropy measures and is positively correlated with randomness or disorder in a physical system.

4   Such modifications relate to the multiple constructions of the commodity and to product differentiation and marketing strategies (see section 4.3).

5   A five-year moratorium on GM crops in the European Union expires in 2003. It seems unlikely to be renewed.

6   Monsanto itself was acquired by Pharmacia in 2000 but then disposed of in 2002.

7   The implications of this for eating practices and consumption are considered further in Chapter 9.

8   Increased growth of non-GM soya has accelerated destruction of the Amazon rain forest, however.

9   Some strands of Marx's writing emphasise people 'conquering', 'dominating' and 'mastering' nature but others display a perceptive understanding of the ecological costs of capitalism and of nature as a source of use values.

10 From one point of view, these laws can be considered as simply cultural constructs. However, since laws such as that of gravity function with regular predictability and have practical utility, it is reasonable to assume that they consistently relate to physical processes.

4 Flows of Knowledge, Circuits of Meaning

4.1 Introduction

In recent years there has been a growing emphasis upon the emergence of a 'new' knowledge-based economy, on knowledge and symbolic products as outputs of as well as inputs to economic processes. There have been claims that knowledge is now the most important economic resource, learning the most important process (Lundvall, 1995). More recently, the emphasis has switched to the centrality of creativity to economic performance, to creativity 'as now powering economic performance' (Florida, 2002, 5). Others have proclaimed the emergence of an economy of signs and spaces (Lash and Urry, 1994). In this chapter I consider these processes of creativity, knowledge creation and flows of knowledge within firms, between firms and between companies and their customers.

A useful distinction can be drawn between three types of knowledge: codified, tacit and 'self-transcending (and not yet embodied)' (Scharmer, 2001, 71). The first two categories are well known, referring to knowledges that have been translated into dis-embodied symbolic forms and those that remain embodied. Recognition of the third, however, draws attention to the 'thought conditions that allow processes and tacit knowledge to evolve in the first place' and as such is important in the context of knowledge creation and creativity. Nonaka et al. (2001, 28–30) identify four types of knowledge assets of firms that build upon the distinction between, and the relations between, codified and tacit knowledges: experiential, conceptual, systemic and routine. Experiential assets are based in shared tacit knowledge. Conceptual assets refer to codified knowledge 'articulated via images, symbols and language'. Systemic assets are those that are codified, systematised and packaged – for example as handbooks or patents. Routine knowledge assets result from the tacit knowledge that is 'routinized and embedded in the actions and practices of the organisation' (ibid., 30). These four types of knowledge in turn form the basis of the processes of knowledge creation and circulation within companies.

4.2 Creativity, knowledge production and flows of knowledge within and between companies

The four-stage SECI process conceptualises the production and circulation of flows of knowledge within firms as involving different combinations of translation and relations between codified[1] and tacit knowledges: Socialisation, Externalisation, Combination and Internalisation (Nonaka et al., 2001). Socialisation involves flows of tacit knowledges. As tacit knowledge is difficult to formalise and is often time-and-space specific, it can only be acquired through shared experience and 'socialisation typically occurs in a traditional apprenticeship'. Externalisation involves the translation of knowledge from tacit to codified form. Codification means that 'knowledge is crystallised, thus allowing it to be shared by others, and it becomes the basis of new knowledge'. Codification is not a simple process, requiring the successful creation of models, languages and messages. Creating models and languages entails high fixed costs but enables messages to be created with low marginal costs. Accomplishing these tasks necessarily requires tacit knowledge. Moreover, typically each step in codification creates new knowledge in the process of making existing, initially tacit, knowledge more widely available. Crucially, however, because the interpretation and use of codified knowledge is *always* filtered through and dependent upon some tacit knowledge, there is scope for a variety of interpretations and the creation of meanings in excess of those intended by the producers of knowledge. Combination involves combining codified knowledge 'into more complicated and systematic sets'. Finally, internalisation is the process of embodying codified knowledge as tacit knowledge and as such is 'closely related to learning-by-doing' (Nonaka et al., 2001, 16–19).

The SECI process underlies the innovative activities in which firms necessarily engage in their search to create competitive advantage grounded in new knowledge that they can, if only temporarily, monopolise. Such innovations take one of three main forms: organisational, product and process. There are strong systemic pressures to find 'new' ways of producing 'old' commodities. Consequently, companies seek process innovations, technologically new ways of making existing products that reduce costs by cutting the labour time needed in production to below the existing socially necessary amount and/or enhance quality. They retain this competitive edge until the 'new' technology diffuses to other producers and becomes generalised, establishing new productivity norms for that commodity (or sector). Knowledge of such innovations can flow via a variety of channels: as texts, via people, via artefacts (Gertler, 2001, 8–12). Product innovation aims to create new markets in which companies can be sole and monopolistic producers. Given the pace and scale of technological change, 'performance superiority will be brief', creating pressures for continuous product innovation (Mitchell, 1998; Wernick et al., 1997, 148). Product innovation can involve creating totally new products or enhancing existing products. It is increasingly important in allowing product differentiation in response to consumer demand in markets for both goods and services (Noteboom, 1999; Poon, 1989). Finally, companies seek to increase productivity via organisational innovation. Within industrial capitalism, this initially revolved around bringing production into factories. Since then, various approaches have been developed to increase the efficiency of production within given technological paradigms (Stalk, 1988). As well as organisational innovation at the immediate point of production, there have been innovations in methods of management (Best, 1990, 35–46).

Nonaka et al. (2001, 20, emphasis in original) insist that 'the movement through the four modes of knowledge conversion forms a *spiral*, not a circle'. In the process of knowledge creation, the interaction between codified and tacit knowledges 'is amplified by each of the four modes of knowledge conversion ... [SECI] is a dynamic process starting at the level of the individual and expanding as it moves through communities of interaction that transcend sectional, departmental, divisional or even organisational boundaries. Organisational knowledge is a never-ending process that up-grades itself continually.' Using its existing knowledge assets, a firm can create new knowledge via the SECI process. This new knowledge in turn becomes folded into the knowledge assets of the firm and, as such, becomes the basis for a new spiral of knowledge creation and for creating the core capabilities of the firm. In brief, it involves the transmission of codified – and often commodified – and tacit knowledges, the former via a variety of media and artefactual forms, the latter involving the movement of knowledge embodied in people.[2]

Clearly, the SECI process is grounded in a particular conception of the firm and intra-firm processes of knowledge creation and circulation, presupposing that, where relevant, the knowledges of *all* workers within the company are drawn upon, and drawn into, these processes. This redefines the spatialities of knowledge creation and flows within companies. Historically, the key privileged sites were the (Taylorist) R&D laboratory, inhabited by men in white coats, or the design studio, the officially designated spaces of creativity. Increasingly, however, there is recognition that *everywhere* within (and indeed outside) the boundaries of the firm is a potential space of creativity and knowledge production.[3] Consequently, knowledge flows within the firm no longer correspond to the hierarchical and linear patterns typical of the routinised Taylorist model of R&D. Within the knowledge-creating firm, knowledge flows horizontally, vertically and diagonally within and between its functional divisions. This circulation of knowledges necessarily depends upon the sharing of codes and languages to allow various communities of interaction and practice (Wenger, 1998) to operate.

As such, the SECI process requires that *kaizen*, continuous improvement through interactive learning and problem solving, generated by an actively committed and engaged workforce that identifies strongly with the company and is dedicated to enhancing corporate performance via co-operating and sharing knowledges, becomes pervasive throughout the firm.[4] SECI assumes the creation of a corporate culture, a common grammar that allows people to make sense of and develop actions in the world, to code history and past experience, informed by a shared 'worldview' and a sense of common corporate purpose. If a co-evolved shared worldview and sense of purpose fail to develop, this model of knowledge creation and innovation is compromised. If they do develop, then they are necessary but not sufficient conditions for successful knowledge creation and corporate competitiveness.

The knowledge-creating firm aims to create and support communities of interaction and build a seamless innovation process grounded in a virtuous spiral of new knowledge creation and transmission. This entails 'management by design' of knowledge creation, learning and innovation via concept teams, limited life project development teams and task forces, for example. Such teams and task forces can smooth knowledge flows and reduce the socially necessary labour time taken to bring new products to market. However, the production of novelty depends upon social interaction within and across the boundaries of these teams and task forces. Consequently, management by design must be blended with self-management by communities of practice.

Increasingly, these teams and task forces are globally distributed, meeting via video-conferencing and other forms of electronic technology. As such, it is increasingly common 'to produce new ideas and products via a communication network that links team members from Singapore and France with those in California and Kentucky' (Leinbach, 2001, 25). Reliance on distanciated social relationships of intellectual production reflects increasing pressures on managerial time and resources but can create problems in transmitting tacit forms of knowledge while working to very tight deadlines (Miller et al., 1996). However, these globally distributed teams do represent a significant change in organising intra-company processes of knowledge creation, transmission and innovation, although Gertler (2001, 19) asserts that 'the idea that organizational or relational proximity is sufficient to transcend the effects of distance (even when assisted by telecommunication or frequent travel) seems improbable'.

Companies also seek to codify tacit knowledge radically to reduce the socially necessary labour time required for the translation of knowledge into innovation. Some 80% of engineering activity is simply minor variation on preexisting practices. By building these routines into computer programs, major companies in aerospace and automotive production can produce new designs in minutes. Generative modelling seeks to capture tacit knowledge and use it further to streamline design and engineering processes. Such a development

prevents knowledge for which [a] company has paid a lot and which differentiates it from its rivals, being lost whenever people leave or die. Engineers, meanwhile, can innovate and create in hitherto impossible ways. What people fail to understand is that very little leading-edge technology is actually new – there have not been too many new laws of physics. If we build in [to computer programs] structured knowledge – knowledge that has immutable laws or inferences – we can go on to innovate in a fraction of the time it took before. (Gareth Evans, Chief Executive of KTI (Knowledge Technologies International), cited in Cole, 1999)

The Taylorist model of R&D lacks feedback loops from users of and customers for innovations to those responsible for producing them. Consequently, in dynamic markets new products may fail while opportunities for others may be missed. In contrast, within the knowledge-creating firm, employees are sensitive to external voices. Familiar examples of the engagement between producer and user in product innovation are the beta-testing of computer software and the involvement of potential users in developing new computer games. Users participate in the knowledge-creation process by using a new product and communicating the results back to its producer. This both diminishes the costs of in-house testing and decreases the distance between software creators and users via establishing information feedback loops. Furthermore, integrating a sub-set of potential customers directly into the product development process helps create demand for the final product (Kenney and Curry, 2001, 51–2).

Synthesising different types of knowledge may therefore produce radical and revolutionary innovations in emergent and unexpected ways via interactive knowledge-creation processes within and across the boundaries of the firm. Indeed, the production and application in production of abstract formal knowledge depends in part upon other sorts of knowledge, tacit skills and capabilities, and trial-and-error behaviour (Arora and Gamborella, 1994, 528). This requires acknowledging the legitimacy and 'voice' of different types of knowledge, not least as radical innovations often challenge the dominant 'logic' within an industry. It necessitates closer integration of R&D with other sections within companies, rather than privileging the scientific knowledge of the R&D laboratory, with far-reaching implications for internal organisation and operation. Such tendencies are observable in flexibly specialised SMEs and in new forms of high-volume flexible production. In its most pronounced and knowledge-intensive form, production becomes a 'design process' and an 'R&D process'; production becomes R&D and the production system operates as an expert system (Lash and Urry, 1994, 96). More generally, creative processes of innovation and learning must be suffused throughout the entire workforce, capturing the knowledge of all workers to enhance productivity and the quality of both product and work in a knowledge-based economy (Florida, 1995)[5]. Flexible production is both innovation-intensive and knowledge-intensive (Lash and Urry, 1994, 121).

Nonaka et al. (2001) argue that knowledge-creation processes occur in a specific time/space, which they denote as *ba* and which they see as pivotal to the process of continuous learning. As such, they emphasise the reciprocal and mutually constitutive relationships between flows and spaces. *Ba* is a relational space, which may be continuous or discontinuous, with fluid and permeable boundaries, capable of rapid change and redefinition by those participating in it. Change therefore occurs both at the micro-level of the individual participants and the macro-level of the collectivity of *ba* itself, in part because the membership of *ba* fluctuates as its members come and go. Nonaka et al. (2001, 24–6) recognise four types of *ba*, a variety of spatialities of knowledge-translation processes: originating, dialoguing, systematising and exercising. Originating *ba* is defined by individual and face-to-face interactions, which give rise to 'care, love, trust, and commitments, which forms the basis for knowledge conversion among individuals'. This emphasises the non-economic grounding of knowledge that becomes an economic asset. Dialoguing *ba* is defined by collective and face-to-face interactions, mainly offering a context for externalisation. In contrast, systematising *ba* is defined by collective and virtual interactions. Technologies such as online networks, groupware, documentation and databases provide a virtual collaborative environment for such interaction. Exercising *ba* mainly offers a context for internalisation as individuals embody codified knowledge.

Knowledge creation within the firm, within *ba*, is thus to a degree a path-dependent process. As such, it has the potential to ossify from a trajectory of knowledge creation and innovation to one of cognitive lock-in, to transform core capabilities into core rigidities. Companies therefore actively seek to avoid this risk. For example, corporate leaders and senior managers may intentionally introduce 'creative chaos to evoke a sense of crisis' (Nonaka et al., 2001, 35). Facing chaos, employees experience a breakdown of familiar routines, habits and cognitive frameworks. Consequently, periodic breakdowns, periods of 'unlearning', provide an important opportunity for them 'to re-consider their fundamental thinking and perspectives.'

Put another way, knowledge-creating companies must ensure that they have the capacity for 'double-loop' as well as 'single-loop' learning (Levinthal, 1996). Single-loop learning involves incremental change within existing production paradigms, and so may entail inertia via cognitive lock-in. Double-loop learning involves radical redefinition of those paradigms, and, as such, an 'unlearning' process, discarding obsolete and misleading knowledge. It reflects the increased grounding of production in discursive knowledge. Knowledge based on reflexivity operates via a double hermeneutic in which the norms, rules and resources of the production process are constantly called into question. This is particularly important in the context of firms characterised by heterarchical organisation. Given such fundamental environmental uncertainty, the critical organisational imperative for such firms is 'the ability radically to question the appropriateness of the assumptions of one's own organisational behaviour. This ability makes for the *reflexivity* of heterarchies' (Grabher, 2001, 354, emphasis in original). Thus production involves complex processes of learning, which vary with firm organisation.

Companies therefore seek to tread the fine line between order and chaos in their ongoing search for creativity and the maintenance of 'requisite variety'. For creativity 'lies in the border between order and chaos' and requisite variety helps a knowledge-creating organisation maintain a balance between order and chaos: 'requisite variety should be a minimum for organisational integration and a maximum for effective adaptation to environmental change' (Nonaka et al., 2001, 36–7). Consequently, the knowledge-creating company must seek to create organisational structures that match internal diversity to the complexity and variety of the environment in which it operates[6]. While innovation has always been critical to corporate practice and self-image, there is now 'much greater attention being paid to fostering the powers of creativity that will foster innovation, most particularly through ... an emphasis on "creative knowledge". Thus creativity becomes a value in itself' (Thrift, 2002, 205),

4.3 Flows of information from companies to potential consumers: advertising and contested meanings

The sale of commodities depends upon flows of information from producers to potential purchasers, both other companies buying commodities as inputs to the production of other commodities and purchasers of commodities for final consumption. Especially in the latter case, advertising plays a critical role in the production and dissemination of knowledge about commodities, seeking to construct conceptual spaces of intended meanings to entice potential purchasers and consumers. Often, however, these intended meanings are contested and challenged, creating instead unintended meanings as a result of consumer resistance and subversion. Producers may respond to this by changing the projected image of the product through advertising or materially altering the commodity that they are trying to sell (Figure 4.1).

Advertising has a venerable history but until quite recently was no more than a marginal influence on patterns of sales and production. In the early stages of the factory system the great bulk of products was sold without extensive advertising. The formation of modern advertising was intimately bound up with the emergence of new forms of monopoly capitalism around the end of the nineteenth and beginning of the twentieth centuries. The development of modern advertising was central to corporate strategies to create, organise and where possible control markets, especially for mass-produced consumer goods. Mass production necessitated mass consumption, and this in turn required a certain homogenisation of consumer tastes for final products. At its limit, this involved seeking to create 'world cultural convergence', to homogenise consumer tastes and engineer a 'convergence of lifestyle, culture and behaviours among consumer segments across the world' (Robins, 1989, 23). Creating mass markets entailed radical changes in the organisation of advertising, both in terms of advertising media and via more conscious and serious attention to the 'psychology of advertising'. The period between 1880 and 1930 saw the full development of an organised system of commercial information and persuasion, as part of the modern distributive system of large-scale capitalism (Williams, 1980).

Advertising, like the products of the industries that it promoted, became organised around the principles of Taylorisation and Fordism, dominated in the

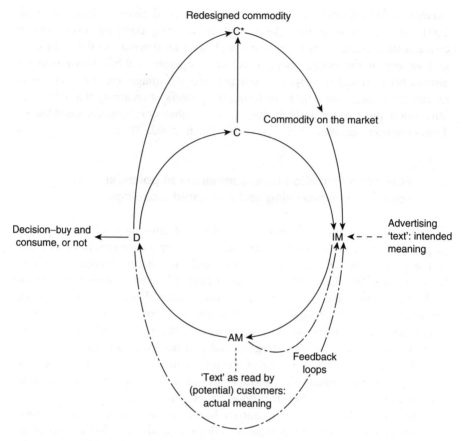

FIGURE 4.1 Advertising and circuits of meaning

USA by a handful of major agencies. They then expanded abroad via processes
of merger and acquisition and foreign direct investment as 'networks' (the term
used by advertisers) of international offices of groups rapidly formed.
Advertising companies drew upon the latest advances in scientific and social
scientific knowledge. Thus in the USA in the 1920s and 1930s the development
of modern advertising drew heavily on current psychological theories about how
to create subjects (Miller, 2002, 174). Initially advertisements concentrated
upon providing 'factual' information about commodities and persuading consumers to buy them on the basis of their modern, sometimes scientific, attributes. Increasingly, however, the emphasis switched to the symbolic connotations
of commodities, not least in the inter-war USA by reference to the images and
symbolism of Hollywood. Subsequently, as consumer psychology further developed in the 1940s and 1950s, it provided the basis for advertising and marketing to take on a 'more clearly psychological tinge' (Miller and Rose, 1997, cited

in Thrift, 1999, 67). In this way and by altering the context in which advertisements appear, they 'can be made to mean "just about anything"' (McFall, 2002, 162) and the 'same' things can be endowed with different intended meanings for different individuals and groups of people. As such the existence of markets requires representations of the consumer (Nixon, 2002, 133).

Indeed, modern capitalism could not function without advertising, which offers mass-produced visions of individualism (Ewen, 1976). Grasping the full significance of advertising requires 'realising that the material object being sold is never enough: this indeed is the crucial cultural quality of its modern forms' (Williams, 1980, 185). Commodities meet both the functional and symbolic needs of consumers. Even commodities providing for the most mundane necessities of daily life must be imbued with symbolic qualities and culturally endowed meanings:

> ... we have a cultural pattern in which the objects are not enough but must be validated, if only in fantasy, by association with personal and social meanings. ... The short description of the pattern we have is *magic*: a highly organised and professional system of magical inducements and satisfactions, functionally very similar to magical systems in simpler societies, but rather strangely co-existent with a highly developed scientific technology. (Williams, 1980, 185, emphasis in original).

Consumers are susceptible to influence via advertising precisely because they have – and, because of the effects of time and space, can *only* have – imperfect and partial knowledge of commodities and markets (Mort, 1997). This creates space for companies actively to seek to change or create consumer tastes and cultivate preferences for new products via advertising rather than simply respond to consumer preferences and demands as expressed through markets. Advertising practice 'constantly problematises the entire notion of "specific products" and constitutes a set of technologies for attempting both to de-stabilise markets and then to re-institutionalise them around new, strategically calculated product definitions'. Advertising operates in an environment in which markets and products are 'continuously and dynamically changing ... advertising focuses on exploiting these environmental conditions, creating variations between product concepts as a means to reconfigure both consumer demand and competitive market structures' (Slater, 2002, 68–73). This power-fully emphasises the way in which advertising practices and products, the latter themselves produced as commodities intended to sell other commodities, can be central to the prosecution of destabilising 'market disturbing' strategies of strong Schumpeterian competition and the redefinition of markets via recreating the intended meanings of commodities.

Barthes (1985, xi–xii) expresses the point as follows: 'Calculating industrial society is obliged to form consumers who don't calculate; if clothing producers and consumers had the same consciousness, clothing would be bought (and produced) at only the very slow rate of its dilapidation'. Consequently, 'in order to blunt the buyer's calculating consciousness, a veil must be drawn around the object – a veil of images, of reasons, of meanings ... in short a simulacrum of the real object must be created, substituting for the slow time of wear a sovereign time free to destroy itself by its own annual act of potlatch'. In fact, the 'annual act of potlatch' has increasingly become one that is enacted several times over the course of a twelve monthly cycle as product life cycles have been shortened and the fashion cycle speeded up. In this way, Barthes opens up a rich symbolic seam that makes it possible to speak of a 'language of fashion' in which clothes function as identity codes for particular cultural and social groups while at the same time creating scope for built-in obsolescence and the creation of continuous demand for clothes produced as commodities (and the point is valid for other commodities). For example, in the USA in the last quarter of the nineteenth century, there was a 'fundamental change' in how people led their lives, a change that convinced most middle-class women that ready-to-wear clothes were better than those they made at home. However, it 'was not sufficient to maintain the constant increase in sales demanded by ... manufacturers and merchants. Sales, therefore were kept high by creating a situation of open-ended demand, in other words by introducing and making available fashion to the middle classes. The dictates of fashion led to perpetual changes in style, and therefore to built-in obsolescence' (Domosh, 1996, 260). Thus the needs of the new department stores for mass sales volumes reinforced the fashion industry, as they fuelled demand by displaying fashion alternatives in settings that imbued those commodities with social meaning. At the same time the department stores, because they sold mass-produced and therefore less expensive items, made available to more women the possibility of being fashionable. This further stimulated production, which led to new demands for consumption. This created a virtuous circle for capital of product innovation and the establishment of meanings for new products via advertising within circuits of capital, commodities and meanings, with the latter critical in so far as the process was driven by 'fashion'.

Subject to the critical qualification that codes of meaning are understood and the differences between items of fashion are marked symbolically, 'a wider realm of signs is communicated that is open to manipulation through advertising, styling, branding and marketing' (Allen, 2002, 56). As such, advertisers are not simply 'choosing' between an array of possible definitions and meanings of products and markets. They are creating and implementing them. Consequently, a marketing strategy 'is not – in the first instance – a matter of competition *within* market structures; rather it is a matter of competition *over* the structures of markets' (Slater, 2002, 68, emphasis in original). Furthermore, as ICTs have developed, providing more potential channels through which to direct messages to potential consumers (television and websites as well as sponsorship and promotional events, for example), choice of medium has assumed greater importance in constructing advertising and marketing campaigns (Nixon, 2002, 137). However, the meanings that advertisers and their clients intend may not necessarily be those that are accepted by intended purchasers and consumers. Consequently, the structures and functioning of markets may also deviate from those intended as markets develop in unanticipated ways.

Furthermore, as Barthes' suggests, 'the economic practices of advertising are intrinsically caught up with the cultural understanding of the role, functions and nature of advertisements' (McFall, 2002, 161). The competitive pressures of contemporary capitalism have stimulated further refinements to 'the magic system', with heightened importance attached to culturally endowed and symbolic meanings of commodities and the identities that people (in part) form through consuming them. There has been a gradual shift in the content and tone of advertisements from simple informational announcements to emotional and symbolic products that seek to persuade. In part, this is because technological advances have presented many consumers with greatly enhanced access to information about products and their attributes, especially as electronic transmission and digitised information have accelerated flows. As a result, few consumer products in almost any category can expect to have any tangible point of difference for very long: 'the USP, or unique selling point, is increasingly a thing of the past' (Stockdale, 2001, 17). In these circumstances the symbolic attributes that can be attached to commodities become increasingly important in product differentiation and meanings.

Baudrillard (1988) argues that stable social relations and practices have disintegrated, leaving space for advertising to capture the pre-existing meanings of things for commercial promotion. In the society of the sign, the capacity to strip objects of their previous meaning has far-reaching consequences. As Featherstone (1991, 66–7) notes of advertising in the post-modern (as he sees it) phase of cultural production, the sign value of things dominates exchange relations and consumer culture. The constant production of artful signs and images in advertising has further blurred boundaries between the everyday and art. Echoing this, Lash and Urry (1994, 4) assert that 'the mobility and velocity of objects in contemporary society have emptied them of their content, so that objects are better understood as sign values rather than as material objects'. They stress the symbolically saturated character of commodities, a consequence of the extraordinary proliferation and circulation of predominantly visual images and signs that become attached to a wide range of commodities. Via such 're-enchantment', mass produced commodities acquire an aura of symbolisation. The semiotic work of advertising therefore involves the skilful deployment of symbols, irrespective of whether the signs themselves represent anything in particular.

4.4 From advertising commodities to managing brands

The de-coupling of signs and symbols from any specific referent product has been further extended with the growing emphasis on promoting brands as opposed to advertising specific commodities. The increasing prevalence of 'enormously powerful and ubiquitous global brands or logos' with a 'fluid-like power' derives from the ways in which 'the most successful corporations over the last two decades have shifted from the actual manufacture of products, which has been increasingly out-sourced, to become brand producers, with enormous marketing, design, sponsorship, public relations and advertising expenditures' (Urry, 2001, 2). As such, it signals a major change in the character of contemporary accumulation. This is the case across a wide spectrum of commodities, both consumer goods for final consumption and intermediate products, even commodities such as metals and chemicals (Harvey, 2003). Brands typically are tied to specific proprietary markers, such as hieroglyphs or logos (for example, the curly script and curvaceous bottle that encourage people all over the world to drink Coca-Cola) or a particular person (such as David Beckham or Richard Branson), which both distinguish the brand and define particular brand families (Klein, 2000). Such logos are deliberately targeted and intended to force the viewer to look at them, 'to underscore the capacity of the brand to condense its message to its mark' (Franklin et al., 2000, 69). This capacity is partly a result of extensive processes of market research and promotion and of the ways in which the phatic inscriptions of the brand create and maintain links among product items, lines and assortments.

Proprietary markers for brands thus operate as phatic images (Virilio, 1991), images that target attention, synthesise perception. As a result, 'the time of the brand is that of the instantaneity of recognition and thus discrimination: brands work through the immediacy of their recognisability' (Lury, 2000, 169). As a phatic image, the brand works to displace or de-contextualise bodily or biographical memory which had been a naturalised aspect of the (apparently instantaneous) processes of perception, and re-contextualise it within its own (technologically mediated) body of expectations, understandings and associations built up through market research, advertising, promotion, sponsorship and the themed use of retail space. Thus practices of thematisation are elaborated and extended in the manipulation of an object's environment, or time/space context. As a result, brand owners frequently present branded objects in themed spaces – parks, restaurants, pubs and shops – or contribute to the elaboration of themed lifestyles through the sponsoring of events or activities. The result is that 'these brands are free to soar, less as the dissemination of goods and services than as collective hallucinations' (Klein, 2000, 22). This creation of such distinct '(hallucinatory) spaces of brands' exemplifies the dialectic between spaces and circuits of meanings.

What is significant about the brand as a phatic image is the extent to which it can 'recoup the effects of the subject or consumer's perception as the outcome of its own powers through an assertion of its ability to motivate the product's meaning and use. This is achieved through the ways the brand operates to link the subject and object in novel ways, making available for appropriation aspects of the experience of product use *as if they were the properties of the brand*'

(Franklin et al., 2000, 68–9, emphases in original). More precisely, this is the intended way in which the potential purchaser should read the brand and be prepared to pay a premium for acquiring it. Purchasers thus pay for the brand name, the aesthetic meaning and cultural capital that this confers, rather than for the use value of the commodity *per se*. These aesthetic and cultural meanings of brands and sub-brands then become ways of segmenting markets by ability to make the premium payments required to possess the desired brand. Successful global brands, such as Benetton, Bodyshop, Gap, Nike and McDonald's have become powerful because they have succeeded in creating 'family resemblances'. Via such commodity kinship, commodities become seen as sharing essential characteristics: 'the shared substance of their brand identities' (Franklin et al., 2000, 69) thus becomes available to those who can and are prepared to pay for the cachet of the brand.

While this de-coupling of brand logo and its meaning from specific commodities may have validity in relation to the brand as a generic representation of a particular company's products, particular objects must nonetheless maintain a degree of stability of meaning so that they can perform as commodities and so enable markets to be (re)produced. The 'overriding assumption' in contemporary marketing techniques is that objects are not defined by their formal properties alone. Customers' needs are seen to give commodities their 'essential character' and products are classified according to the way that they are perceived, bought and consumed, resulting in product categories such as 'convenience', 'speciality' and 'unsought'. These in turn may become the basis of brand portfolios. In this way, the properties of objects 'are linked to their positioning relative to customers' perceptions and needs, *but only as documented, interpreted and re-presented by the advertising or marketing industries*' (Lury, 2000, 168, emphasis in original). Consequently, there are limits to such a process of signification, which vary with the types of object and their use values. In contrast to Baudrillard and Lash and Urry, others (for example, see Appadurai, 1986) have demonstrated that the meaning of objects derives from their uses, forms and patterns of circulation. 'Meaning, then, might be more accurately conceptualised as at once a semantic and pragmatic category ... use cannot be easily separated from meaning precisely because use is itself a major factor in the determination of meaning' (McFall, 2002, 152). Thus the use of commodities is intimately linked to their circuit of meaning; and vice versa.

Nevertheless, not withstanding this important qualification, the symbolic aspects of many commodities have clearly increased in significance. Consequently, people's identities have been 'welded to the consumption of goods' (Ewen, 1988, 60), and indeed to particular brands, linked to further product and process innovations in the form of advertisements. One corollary of the enhanced significance of cultural capital and the meanings of things has been a deepening social division of labour incorporating a complex and sophisticated advertising sector in addition to advertising divisions within companies. The growing significance of advertising and marketing has generated 'white-collar' jobs, heavily concentrated in major urban areas. Advertising knowledge and skills have become critical resources and marketable commodities. Moreover, in strong opposition to those who argue the case for 'consumer sovereignty', Williams (1980, 193) emphasises that "in economic terms, the fantasy operates to project the production decisions of the major corporations as "your" choice, the "consumer's" selection of priorities, methods and style'. With the growth of department stores in the nineteenth century, shopping became a skilled, knowledge-based activity, but with consumers' knowledge controlled by retailers and advertisers (Glennie and Thrift, 1996b, 224–5). Or as McDowell (1994, 160) puts it: the production, advertising and marketing of goods is a crucial part of their consumption, 'as anyone who wears Levis knows!' This highlights the ways in which companies seek to imbue commodities with particular meanings via advertising and marketing strategies.

Others have also emphasised the increased significance of advertising for contemporary capitalism in creating demand via appeals to consumer individuality and identity. According to Jameson (1988, 84), there is now an absolute pre-eminence of the commodity form – the logic of the commodity has reached its apotheosis. This is based upon a heterogeneous market that thrives on difference and incommensurability, fuelled by the cut and thrust of 'symbolic rivalry, of the needs of self-construction through acquisition (mostly in commodity form) of distinction and difference, of the search for approval through lifestyle and symbolic membership'. This elaboration of commodification has been made possible by the development of the mass media, especially television. Advertising and design companies have produced an enormous machine for generating a desire for commodities – a greatly enhanced 'magic system' – via more powerful, sophisticated and persuasive processes of sign production. Such production via the advertising industry is a necessary condition *for* exchange relations and the circulation of capital, the dominant driving process.[7]

Some argue that such a perspective tends to overemphasise the power that advertisers, allied to retailers, can exert over consumers (Jackson, 1993). Advertising is rarely the sole or even most important source of pre-purchase knowledge about the existence or qualities of particular commodities, 'seldom the single stimulator of wants and desires' (Pred, 1996, 13). It is certainly the case that 'consumers do not *straightforwardly* draw upon meanings prescribed by retailers and advertisers, but rather that commodity meanings are often contested and re-worked by consumers' (Leslie and Reimer, 1999, 433, emphasis added). While producers 'create a series of texts', these are 'read by different audiences according to their own social conditions and lived cultures' (Jackson and Taylor, 1996, 365). While the process may not be straightforward, advertising undoubtedly can exert enormous influence in mediating and shaping relationships between the sign values of commodities, their symbolic meanings, and their material content and form (Fine and Leopold, 1993, 28). One can, however, go too far in celebrating consumer autonomy, reflexivity and resistance. Not least, companies continue to realise surplus-value via the successful sale of commodities, suggesting that advertising strategies have considerable efficacy in relation to reproducing capital on an expanded scale. While Jameson may underestimate the capacity of consumers to challenge or even subvert the commodification strategies of capital (Thrift, 1994, 222), it is more difficult to challenge claims as to the creation of demand via advertising and marketing strategies.

However, McFall (2002, 148–9) correctly cautions against a too ready acceptance of binary epochal accounts that posit a sharp shift from an era in which advertising was informational to one in which it is persuasive. The form that advertising takes at any given moment, and the meanings and messages that it is intended to convey, are context dependent, a product of the interplay of a specific conjuncture of forces and processes. As such, there is a need for a lower level of analysis that encompasses the institutional, organisational and techno-logical specificities of the production of advertisements and the particular forms that they take, rather than simply appealing to epochal accounts and claims.

4.5 Circuits of meaning, advertising strategies and the co-production of meanings

The concept of 'circuits of culture' suggests a nuanced view of the creation, transmission and receipt of meanings, allowing for recursive loops, feedback from consumers to producers, and learning by producers from consumers (Johnson, 1986).[8] The creation of meanings is a continuous process. The start-ing point is the creation, within given social conditions, of a series of texts by producers, which are then read and interpreted by different audiences accord-ing to their social conditions, positionality and lived cultures. Both the inten-tions of brand managers and the attributes and positionality of potential consumers shape reactions to adverts. Indeed, some of the more effective TV adverts are those that require potential consumers to 'work with them', actively to engage with them so that viewers have to supply part of the meaning themselves (Jackson and Taylor, 1996, 360). Audiences' culturally constructed knowledges therefore play a key role in the decoding and interpretation of media messages and the ways in which adverts are understood as they under-take the cultural work of interpretation – and such understandings are likely to vary significantly between time/spaces and types of people. Moreover, this process of decoding is not simply a semiotic process but also involves the uses that people make of things and use is a major factor in the determination of meaning.

According to Lash and Urry (1994, 277), consumers have become less susceptible to the illusions of mass consumption. This is certainly true with respect to specific spaces and social groups, though not universally.[9] However, the process of consumers chronically challenging and reworking the meanings of commodities that they have purchased has wider implications. Conceptions of globalisation postulating the creation of homogeneous global markets are untenable and indeed miss the point. The recognition that the ways in which advertisements are 'read' is culturally constructed and varies over time/space, and with the class, gender, ethnicity, age and so on of the 'reader', allows companies to use advertising strategies to segment markets by seeking to create meanings that are specific to these segments. In this sense, advertising is an inherently spatial practice, creating and differentiating circuits and spaces of meaning. One has only to consider the changes to the advertising strategies for Coca-Cola to appreciate the point. After decades of a strategy based on the message of 'one sight, one sound, one sell', Coca-Cola has sought to devise an advertising strategy that seeks to respond to local specificities and 'to make Coca-Cola appeal to every type of consumer, of every culture and nation, on every occasion' (Mitchell, 1995, 61, cited in Jackson and Taylor, 1996, 364). This exemplifies the way in which major multinationals are increasingly acting globally but thinking locally. Increasingly, they are devolving responsibility to local branches or agencies for creating adverts that are customised to local conditions – variations on a global theme, but tailor-made to fit local circumstances, increasingly multilocal rather than variations on a multinational theme.

From a starting point in the intended messages of advertisers, typically targeted at a specific market segment or niche, circuits of meaning continue through successive phases of consumption and production. Consumers 'second guess' advertisers' intentions while producers vary the nature of the product in response to or in an attempt to anticipate consumers' reactions. Advertisers are in constant contact with consumers as they seek to monitor and anticipate reactions, drawing on a range of social science disciplines, skills and social research methods. Such information is used to evaluate existing campaigns and in turn deployed to structure the form and content of future advertising campaigns and possibly to lead to product innovations as the commodity, and not just the ways in which it is represented, is altered (Figure 4.1).

This emphasises the ways in which innovations in the form of advertisements result from the interplay between their producers and consumers. Grabher (2001, 352) comments that:

> The production [of] the 'first wave' advertising by US multinational agencies was more utilitarian, focusing on the functional attributes of commodities. ... In contrast, the second wave seeks to confront the growing opposition, scepticism and resistance on the part of the consumer by creating a new type of advert inspired by irony, self-deprecation and self-reflexivity. The aesthetic of the second wave opposed the bombastic, declaratory, or literal style of the first wave with unusual and subtle visual presentations. Pioneering London agencies in this breakthrough from the first wave systematised this departure in product innovation through a new direction in process innovation. That is, an advert is 'planned' for an account by testing it out on small samples of consumers.[10]

Using focus groups as part of the design process enables the views of potential purchasers and consumers of a commodity to influence, often decisively, the shape and final form of the advert. Subsequently, this emphasis on creativity was taken further in a third wave of advertising. Grabher goes on to argue that 'with the challenge of market-research driven advertising by a new ethos of creativity, the hegemony of the major US agencies has been broken. Soho, the epicentre of the second wave, to re-phrase the sea-change in terms of the geography of production, had emancipated itself from Madison Avenue'.

Consequently, advertising is becoming more like a culture industry (Lash and Urry, 1994, 139). Advertising firms typically have two functions – to make and to place advertisements. As such, 'they have a "creative" side, and a more purely business side, which is their function as "media space brokers"'. Echoing this, Grabher further claims that account planning is emblematic of the implosion of the economic, of the increasingly cultural inflection of advertising as a business service, and of the transformation of advertising into a 'communications' or 'culture' industry. While there may have been such an implosion, it remains governed by strict commercial criteria and marked by precise economic limits since advertising agencies need to make profits and the adverts they produce need to sell other commodities. At the same time, therefore, this growing 'culturalisation' of advertising is not without economic intent and effects. There is an increasing tendency for public discourse about advertising to expand beyond the narrow boundaries of the business community to comments on and reviews of advertising campaigns in daily newspapers and popular magazines and television programmes that seek to create popular entertainment from adverts, such as *Jo Brand's Commercial Break*. Although the direct impacts that this expansion and popularisation of discourses have on particular advertising styles or philosophies may be marginal, '*it reinforces a feedback loop* that supports a continuous up-grading of the industry. Broadening the debate leads to a deepening of the "advertising literacy" of the consumers who, in turn, make increasing demands on the sophistication and subtlety of advertising' (Grabher, 2001, 370, emphasis added). This points to the ways in which the forms and content of adverts themselves and their meanings are reworked within a circuit that encompasses diverse producers and consumers.

Such circuits, however, can include a variety of consumers but exclude producers, presenting difficulties for advertisers and brand managers as a result. This is indicative of the limits of focus group approaches and second and third generation advertisements and of the need to deploy new methodological approaches in understanding circuits of meanings. As with most things in popular culture, people's understandings of brands come from an oral tradition, 'from the passing of stories. Consumers in effect gather around metaphorical camp fires – websites, newspapers, the pub, their kitchen table – swapping and listening to stories about brands. And it is these stories that drive brand meaning' (Stockdale, 2001). The problem that this poses for brand owners is that 'they are not there'; they are excluded from these spaces. As a result, all points of contact with consumers must now be seen as opportunities for learning and R&D, rather than as simply points of sale or complaints to be avoided. Consumers need to be engaged, their views on products understood and used to enhance product perceptions and market shares. Indeed, 'disaffected consumers' could be recruited to help find solutions to problems, or help develop new products. It is 'only' by using such techniques that 'marketing people can get back into the game and start taking charge once again of what their consumers think of their brands. One of the biggest challenges is to understand the implications for the role of advertising and communication in all this – especially as we move towards a more interactive media world where viewers can simply elect not to see advertising or sponsorship messages' (ibid). Whether Stockdale's proposals would indeed eliminate the managerial problems of consumers challenging the intended meanings of adverts, ensuring that they read them in the 'intended' way, is debatable (and unlikely). However, it does indicate the complexities in understanding flows of meanings and the ways that these can be transformed as they pass around various socio-technical circuits.

Advertising often offers a range of 'uneasy pleasures', from the simple desire for a product and its associated 'lifestyle' to more complex pleasures such as the enjoyment of an 'in-joke' and other forms of audience participation (Jackson and Taylor, 1996, 360–1). Advertisements seek to enhance these pleasures through the use of complex narrative structures such as mini soap operas and inter-textual references to previous well-known adverts. However, people are bombarded with a constant stream of images, one expression of intensifying time/space compression, a manufactured diversity that they (or at least some of them) have become skilled at interpreting by performing semiotic work. Consequently, 'the symbolic interplay that constitutes consumer codes is thus not something that is handed down through a marketing tradition but is itself open to manipulation by active consumers' (Allen, 2002, 41). As such, Lash and Urry (1994, 277) claim that people are increasingly reflexive about their society, its products and its images, 'albeit images which are themselves part of what one might term a semiotic society'. This raises critical questions as to *which* people have the capacity to become 'active consumers'. While this claim may have validity in some socio-spatial circumstances – that fraction of the new middle class endowed with ample cultural capital and writing about itself? – there are evident dangers of overgeneralisation here. There is, for example, little evidence of people becoming 'active consumers' over much of the marginalised spaces of Europe and North America, let alone sub-Saharan Africa.

The pleasures to which Jackson and Taylor allude are uneasy because of consumers' knowledge that advertising is designed to sell as well as educate or entertain, that advertising effectively masks the social relations of production, and that their independence as consumers may be subtly undermined by the incorporation of consumer resistance. One expression of this is the growing resistance to brands, famously expressed in the slogan 'No Logo' (Klein, 2000). Increasingly, advertisers seek to utilise a cultural politics of irony, saying one thing while meaning another, and inviting consumers to play with the sense of 'double meaning' that advertisements frequently evoke. For irony is always double-edged, capable of expressing resistance to dominant readings but always liable to appropriation and incorporation. It is this range of meanings that the notion of 'uneasy pleasure' conveys. It suggests that the circuits of meaning associated with advertising and the interplay between producers and consumers of meaning has become complex and, by design, loaded with ambiguity and scope for alternative and contested readings.

4.6 Circuits of alternative meanings: new meanings, re-valorising commodities

Consumers can not only challenge the original intended meanings of advertisements but can subsequently create alternatives to them, endowing 'old' things with 'new' meanings and in the process re-valorising them. As commodities reach the end of their useful life for their original purchasers or recipients, they may donate them to gift or charity shops, or sell them in a range of informal spaces of sale, which imbue these 'second-hand' or recycled commodities with fresh meanings. Their new purchasers may further rework the original dominant or intended meanings of things in the creation of secondary (and maybe then tertiary, quaternary and so on) circuits of meanings. This problematises the claim that commodities have only one socially necessary meaning and existence.

Rituals such as 'gift giving' are another social process through which meanings are (re)defined, becoming 'culturally drenched' and taking on 'identity value' (Featherstone, 1991). Although there is a well-established tradition of analysing gift exchanges in non-monetary, non-capitalist and non-western economies, there is an assumption that the exchange systems of the capitalist world are thinner, less loaded with social meaning and less symbolic. However, this assumption is contentious, particularly in the context of gift giving. Motivated by concerns such as affection, love and friendship, gift giving is an issue of great economic and symbolic significance in the monetised economies of capitalism. This was the case in the era of nineteenth-century modernity and remains so in early twenty-first-century late modernity, especially in terms of the significance of rituals such as birthdays, name days and Christmas (Domosh, 1996). A plethora of commodities – cars, clothing, perfumes, toys and so on – are purchased with the intention of them being given as gifts rather than consumed by their purchaser.

The affective, emotional and symbolic aspects of exchange, in which interpersonal gift giving is grounded, are important determinants of transformations in meaning (Crang, 1996, 50). Such gift giving is both a form of exchange and of social communication (Strathern, 1988), with the gift simultaneously expressing altruism and egoism. Bourdieu (1987, 231) deftly catches the ambiguity of the gift. On the one hand, it is experienced (or at least intended) as a denial of self-interest and egoistic calculation, an exaltation of generosity, a gratuitous gift that there was no requirement to give; on the other hand, it never entirely excludes its constraining and costly character. As such, it is inherently ambiguous. The character of 'costliness' is itself variable, however. It may simply reflect the monetary cost of the commodity, acquired to be subsequently given as a gift. However, it may also reflect the commitment of time and effort required to search for such a commodity in spaces of sale.

In other cases, things are never commodified in the first place, with their meanings derived from a variety of non-commodified relationships. For example, things may be produced specifically as gifts to be given to others. The effort and time required in order to produce something specifically as a gift for a particular person is an expression of affection, love or friendship that cannot be represented through the product of a monetary exchange in the market. Another example is that of production in the social economy, with the intention of providing non-commodified socially useful goods and services. In this case too, the production process is grounded in non-monetised value systems.

4.7 Summary and conclusions

In recent years there has been a resurgence of interest in the knowledge base of the economy, expressed in claims as to the emergence of knowledge-based economies. Circuits of knowledge are certainly of central importance in the performance of economies (and always have been). They involve a variety of flows, within firms, between firms, between producers and consumers, and between private sector and public sector organisations. This reliance upon knowledge flows emphasises that the production of the economy is simultaneously discursive and material, and in recent years there has been a growth in symbolic products as well as heightened attention to knowledge as an input to commodity production. While flows of knowledge within and increasingly between firms are critical to the production of commodities, the sale of commodities to consumers depends heavily upon the meanings that those things come to have. Advertising of particular products and the creation and promotion of brands are therefore pivotal in realising the surplus-value embodied in commodities and so in helping assure the smooth flow of value and expansion of capital. However, because consumers are active and knowledgeable subjects, they do not necessarily absorb the meanings intended via advertising and brand promotion but seek to contest and challenge them in various ways. This may involve resistance to particular commodities or brands, or more generally to any brands – expressed in the slogan 'No Logo' – or seeking to give new meanings to old commodities as these reach the end of their intial socially-useful life and of their first circuit of meaning, setting in motion a second (or third, fourth and so on) circuit. In so far as this recycling of things depresses demand for new commodities, it has an impact upon the circulation of capital and the possibilities for and pace of accumulation, while possibly opening space for consideration of alternative economic logics more linked to concepts of sustainability.

Notes

1   In fact Nonaka et al. refer to 'explicit' rather than 'codified' knowledge.
2   The idea of the 'learning firm' has recently been expanded to encompass different aspects and ways of learning: by doing, by interacting, by using and by searching (Hudson, 1999).
3   Spaces and spatialities of knowledge production and innovation are discussed further in Chapter 7.
4   To shopfloor workers, however, *kaizen* often appears differently (see section 8.2.3).
5   See also section 7.2 for discussion of these issues.
6   Nonaka et al. (2001, 37) suggest that companies can realise requisite variety by developing a flat, flexible organisational structure (see Chapter 7) and/or frequent changes to their organisational structure and/or the frequent rotation of employees (see Chapter 5).
7   In contrast, Featherstone (1991, 66–7) argues that in contemporary (post-modern) cultural production, the sign value of things dominates exchange relations and consumer culture.
8   See also sections 4.2 and 7.5 on the recursive character of innovation.
9   See Chapters 8 and 9 and the discussion of 'post-shoppers' and 'post-tourists'.
10 This also registers a significant shift in social science methodologies, from the large-scale questionnaire surveys to ethnographies, focus groups and semi-structured interviews.

5 Flows of People

5.1 Introduction

Complex economies, based around deep social and technical divisions of labour and spatial separation of spaces of residence, consumption, exchange and production, are characterised by large-scale flows of people between these locations. While flows of people in space/time vary in terms of volume, distance, mode of transport used, frequency, rhythms, and length of stay (and the same person may be involved in flows of varying spatialities and temporalities), there are nevertheless definite regularities and patterns in these flows. These spatio-temporal regularities are a consequence of the performative and social character of the economy, which necessitates that its practices generally need to be carried out collectively in specific time/space settings because of such coupling constraints (Hagerstrand, 1975). These varied flows of people have been made possible by technological innovations in transport, of which the automobile is the most influential (Urry, 1999), which themselves have provided major opportunities for capital accumulation. There are powerful arguments that processes of time/space compression and convergence are facilitating faster, more frequent and more distant flows (Harvey, 1989). For example, in 1950 there were 25 million international passenger arrivals; currently there are over 600 million each year, and the scale of future movements is forecast to increase markedly. The number of international migrants has more than doubled over the last 35 years, to 175 million in 2000 (International Organisation for Migration, 2003). Even so, the vast majority of flows of people remain of a relatively short distance, within rather than across national boundaries.

The collective and social character of economic processes requires that the 'right' people (or more precisely people in their 'right' roles as consumers, sellers or producers) be in the 'right' time/space to enable successful performance of the economy. As a result, flows of people must be co-ordinated and managed. This in turn requires appropriate material infrastructures – trains and boats and planes, often produced as commodities, and their respective roads, railway stations, signals and tracks, harbours, airports and air traffic control systems to allow people to move by various modes of transport. More often than not, the (national) state plays a key role in the provision of such infrastructure, linked to the diversion of capital into secondary circuits. Several of these spaces (such as airports) are 'deterritorialised', examples of 'hyperspaces'. Spaces such as airport transit lounges are neither spaces of arrival nor of departure but 'pauses' in journeys, consecrated to circulation and movement (Urry, 1999, 14). Moreover, transport vehicles such as aeroplanes and ships are 'mobile hyperspaces' (Sampson, 2003), connecting various nodes in networks of mobility, containing large numbers of people in transit. For example, at any one time 300,000 passengers are in flight above the USA while automobiles constitute mobile semi-privatised capsules for movement (Urry, 1999, 13). Such vehicles are mobile elements that stand for 'the shifting spaces in between the fixed spaces' that they connect (Gilroy, 1993, 16–17).

These commuting and migratory movements are part of broader and more generalised patterns of human mobility, as people also move as shoppers, in pursuit of education or leisure and recreation. These movements vary in spatial scale and temporal reach (Figure 5.1). Movements vary in spatial extent from the local to the global. There is a persistent marked distance decay effect associated with many types of movement. There are various reasons for this, including: the transport costs of moving increase with distance; longer moves require increased time and so greater foregone earnings and intervening opportunities; the increased psychic costs of separation from home; and the greater costs of acquiring information about more distant places. Even so, Lucas (2001, 323) argues that 'it is remarkable that we know so little about the underlying causes behind the deterrent effect of distance'. Movements also vary in temporal duration, from hours to years. In general, however, there has been a shift from longer-term to shorter-term mobility (facilitated by the increasing ease and falling costs of travel) and to increasingly multi-purpose movements, combining elements of both consumption and production across a range of spatial scales (Williams et al., 2004). New forms of mobility can be found at many spatial scales, especially for young, single adults and the active elderly, groups who are generally unconstrained by the necessity to conform to the regular rhythms of commuting and being at work. There are 'new forms of mobility which were unimaginable a generation earlier' and which cut across conventional categorisations of movements. For example, 'the young Asian working in New York to pay for a graduate course may simultaneously be a student, a labour migrant and a tourist' (Williams and Hall, 2003, 2).

5.2 Temporary flows linked to the performances of people in the economy

5.2.1 Daily commuting

Daily commuting is a taken-for-granted feature of everyday life over much of the late modern world. Prior to the rise of industrial capitalism, however, the separation

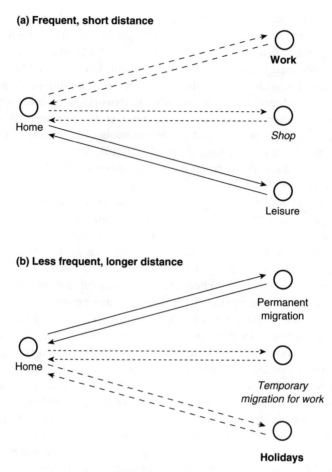

FIGURE 5.1 Flows of people in economies

of spaces of residence and work was unusual. Indeed, this is still the case in many non-mainstream and non-capitalist forms of economic activity and in certain respects homeworking has remained, and even in specific instances increased somewhat, within the social relations of capitalist production.

The origins of routine regular commuting are therefore to be found in the establishment of capitalist relations of production and in particular in the growth of the factory system. This system, innovative and revolutionary at the time of its invention, reflected the need to bring workers within the same time/space to ensure control over the labour process, discipline at work and smooth, uninterrupted and profitable production. As individual spaces of work increased in size, often employing thousands and sometimes tens of thousands of workers, the magnitude of such flows increased correspondingly. Often increased volumes of commuting were directly linked to burgeoning suburbanisation and the spatial expansion of the urban built form, made possible by improvements in and/or spatial extensions to mass public transport systems and the emergence of mass private car ownership. Mass car-based suburbanisation began in the USA the 1920s and 1930s, continued in the UK in the 1950s and subsequently spread to continental Europe and beyond. Typically, the state took on the costs of providing such public transport systems and of providing the investment in road systems to allow the expansion of private car ownership. Initially such commuting movements were constrained by distance and/or the availability of public transport, limited to particular segments of cities. Reliance upon public transport systems typically led to radial flows converging on city centres, a massive surge in the morning and out again in the afternoon or early evening.

At different times in different parts of the world, the variety of spatial patterns has increased and dominant spatial patterns have altered. As availability of the private car and alternative means of private transport increased and geographies of employment and the preferences of those able to exercise choice of residential areas altered, so too have commuting patterns changed. For example, there has been growth in commuting flows from the centre to the suburbs, and between suburban locations around the edge of the city, reflecting the resurgence of city-centre living and the growth of edge cities (MacLeod, 2003). Such tendencies were perhaps most manifest in the USA in the latter part of the twentieth century. For example, there was a rapid growth in reverse commuting from city centres to suburbs and from cities to non-city locations, as well as suburb to suburb and suburb to non-city locations (Glaeser et al., 2001, 33–5). As transport technologies have improved, those who can afford to do so have increasingly chosen to commute long distances between places of work and residence, to live in environments seen as residentially more desirable and/or substantially cheaper. For example, in the UK there are (in 2003) more than 100 daily rail commuters from York to London, a daily return trip of around 750 kilometres, paying over £8,000 for an annual season ticket (in 1993 there were six: Webster, 2003). Over 1,000 commute daily by rail between Peterborough and London, a round trip of over 300 kilometres, while increasing numbers of people are spending between five and seven hours a day commuting to London (Pickard, 2002). The end result of this interaction between labour and housing markets is a complex geography of daily flows into, out of, around and between cities and towns as people travel from their home to work and back again, typically combining rail and road travel.

As well as the distances of daily commuting stretching further, in some ways the strict temporal regularity of these patterns has also become loosened, albeit at the margins. For example, the growth of 'flexi-time' has stretched the length of the commuting period. Not withstanding the increase in 'flexi-time' working, for many activities there are quite definite times in the diurnal cycle between which people must be at work, but for other activities flows are of a much more around-the-clock character. For example, as many major food retailing stores in the UK have opened on a 24-hour a day, almost seven days a week basis, there are flows of workers (as well as shoppers: see below) to and from a variety of typically out-of-town locations at regular intervals. Commuting flows to and from public and private services such as hospitals also necessarily occur around the clock. Other and very different types of work equally require such flows. The production of adverts in Soho, typically to a very tight just-in-time schedule, requires a certain type of culture of work and associated forms of commuting: 'particular place-based conventions with regard to the organisation of work are seen as essential pre-conditions for a cyclical project-based production process. In addition, quintessentially cosmopolitan features, such as 24-hour-and-7-day-a-week availability of key services, facilitate this type of cyclical work regime' (Grabher, 2001, 367). People journey to and from work around-the-clock in flows that are irregularly distributed over time in contrast to the routine of arrival and departure at fixed times.

The spatial regularities of daily commuting have also been further loosened because of innovations in the organisation of work. First, a growing number of people work partly from home, and as a result no longer commute daily. This partial reunification of spaces of residence and work reflects changes in ICTs as well as spatial variations in labour market conditions. Secondly, a growing number of people whose work is increasingly peripatetic no longer have regular commuting patterns, as they do not have a fixed and single work space. They work in a variety of spaces, a mobile mixture of airport departure lounges, aeroplanes, trains and hotel bedrooms, as well as corporate offices in various locations. Thirdly, the spatial regularity of commuting patterns has been disturbed by the growth of 'flexibilisation' and contract work, especially in the USA, the UK and a few other European countries such as the Netherlands and in a range of lower-level service sector occupations. This has been linked to the growth of temporary staffing agencies, to which other companies sub-contract recruitment. Individuals register with these agencies and are then moved from company to company, shifting spaces of work. As such, their commuting patterns become much more variable as spaces of employment alter abruptly.

5.2.2   *Longer-term but still temporary movements for work*

Daily commuting movements co-exist alongside a range of other types of movement, of varying durations and distances. First, commuting can occur on a weekly (or longer) as opposed to a daily basis, often linked to uneven development and differences in economic well-being and prosperity at a regional scale. For example, the most publicised and statistically well-documented spatial cleavage in the UK in the 1980s was the inter-regional 'North–South Divide'. During the 1980s there was a burgeoning gap between these two parts of the UK. One expression of this was the growth in long-distance weekly commuting, as the 'Tebbit Specials', the early Monday morning and late Friday evening trains that brought workers from places such as Liverpool and Middlesbrough to and from the London job market.[1] These were (mainly) men who could not find work in the cities and regions in which they lived but could not afford to migrate to London or elsewhere in the south east because of massive regional house price differentials. They therefore became long-distance commuters, living in the cheapest possible lodgings or sleeping on-site in London. It has been estimated that there were some 10,000 such 'industrial gypsies' (Hogarth and Daniel, 1989), although the phenomenon of long-distance long-term commuting is by no means confined to the UK in the 1980s.

Secondly, seasonal migration is another form of extended and peripatetic movement in search of work. There is a long history of seasonal movement, following the natural rhythms of agricultural production and cropping patterns, especially in relation to the labour-intensive work of harvesting. Such movements of people in pursuit of work are widespread, and often large-scale, especially in parts of the global economy in which agriculture remains a major sector of employment. For example, some 60,000 seasonal migrant workers are employed to pick strawberries in the Spanish province of Huelva (Wagstyl, 2004) while almost 20% of people migrating in Thailand in the late 1980s and early 1990s were reportedly seasonal migrants (Lucas, 2001, 324). In regions of agricultural production in or near to the core territories of global capitalism, seasonal agricultural work is often performed by illegal migrant workers – for example, Albanians in Greece, Mexicans in California or people from various parts of the Balkans in the UK. Such workers occupy very vulnerable niches in labour markets, precisely because they lack legal status and citizenship rights.

Thirdly, there are international labour migration systems, moving people simply to exploit their labour-power for varying periods of time. Often, historically, this was to locations not of their choice. The forced movement of an estimated 6 million African people as slaves to the Americas as part of the transatlantic 'triangular trade' between around 1500 and 1800 was central in establishing the capitalist economy in Europe and (what became) the USA. As the African slave trade declined, a new stream of migrant workers moved between continents: some 30 million Chinese and Indian indentured labourers ('coolies') were recruited, often forcibly, to work in other parts of the Dutch and British colonial empires (King, 1995, 11–14). The opening up of these colonies and the growth of the USA created new spaces for massive permanent migrations of people in the nineteenth and twentieth centuries (which are discussed below). However, the second half of the twentieth century also saw the emergence of new temporary international labour migration systems. Initially, as restrictions were imposed upon permanent immigration, this was to help meet burgeoning labour-power demands, especially for unskilled manual labour, in the core territories of the capitalist economy, notably the USA and western Europe. Subsequently, they focused on more localised 'hot spots' of growth such as the Middle East and Gulf states linked to the production of oil and natural gas.

These, very largely male, migrant workers mainly originate in spaces on the periphery of the capitalist economy. For example, those in western Europe were overwhelmingly drawn from its southern Mediterranean and north African fringes, although an increasing number arrived from south-east Asia in the latter part of the twentieth century; some 66% of those in the USA and over 90% of those in the Gulf states are drawn from peripheral countries (in the latter case, for cultural/religious reasons, predominantly other Muslim countries). These international labour migration movements have become more generalised, drawing in a greater range of countries as origins, with growing differentiation in types of migrant and economic motives for migration, both in terms of the demands of the destination areas and the pressures giving rise to migration in the origin areas. However, while the numbers of international migrant workers have increased to perhaps 30 million, 'the vast majority of workers have never worked outside their country of origin' (Williams and Hall, 2003, 19). This suggests that there are definite limits to the expansion of international migrant labour for cultural, social and political, as well as economic, reasons and that it is important to keep it in perspective relative to other scales of labour mobility.[2]

While formally voluntary movements of people in search of work, national states were typically heavily involved in constructing, managing and regulating these temporary migratory systems, defining the terms and conditions on which international migrant workers were permitted to be present for a specified period of time. There is typically a periodicity to such movements, linked to fluctuations in the labour market as demand for labour rises and falls. Faced with labour shortages that threatened to slow the pace of accumulation and national economic growth and endanger corporate profits, states and companies sought to recreate favourable labour market conditions *in situ* via temporary labour migration, using migrant workers to ease labour shortages. As these demands moderated, and even more as problems of unemployment replaced those of labour shortages in national economies, pressures to stem the flows of migrant workers, and repatriate those already present, often grew. For example, in western Europe the period of active recruitment in the 1950s and early 1960s was followed by one of selective closure, especially for unskilled workers. Germany can be taken as the emblematic European example. The erection of the Berlin Wall in 1961 stemmed the supply of cheap labour from Poland and the German Democratic Republic to West Germany. In response, the German state, in co-operation with major companies, sought to fill the gap in the labour market by establishing a variety of international labour migration schemes with states in Mediterranean Europe and North Africa. This resulted in the particularly rapid growth of 'guest workers' (*gastarbeiter*) in the 1960s, to over 10% of the German labour force, and they became structurally embedded in occupations that ethnic Germans were reluctant to fill. As unemployment subsequently rose from the later 1960s, the flow of new migrant workers slowed and pressures to reduce the stock of such workers in Germany grew. Most temporary migrant workers lacked citizenship status and were employed on specific fixed-term contracts (although this is now beginning to change, not least as it became clear that they did jobs that others would not do).

More recently, national governments, not least that of Germany, have sought to facilitate the recruitment of skilled workers, or even themselves actively to recruit such workers, in key sectors of national labour markets facing acute shortages of skilled labour, by introducing a variety of regulatory and taxation measures (Maclaughlan and Salt, 2002). In addition, private sector temporary employment agencies have become involved in organising such international migration flows, across a range of occupations but especially in response to growing demands for skilled international migrant workers and certain types of unskilled worker. This has, to a degree, led to a feminisation of international labour migration flows, with women occupying a leading role in many migration streams. For example, Cape Verdian, Filipino and South American women have migrated to work as temporary domestic helpers and carers in parts of Europe (King, 1995, 23), participants in global care chains as care work is commodified and internationalised (Lutz, 2002).

However, while state and private sector organisations may orchestrate such flows, the precise patterns of origins and destinations often involve friendship, family and household relations. These family and social networks link particular people from urban neighbourhoods or villages in origin countries with particular workplaces in destination countries within the broad currents of the macro-flows. Furthermore, as the regulation of unskilled labour migration has tightened, pressures for irregular and illegal migration have grown, sometimes linked to asylum seeking as a result of geopolitical instabilities and war. As the formal regulatory system has become more restrictive, the importance of informal networks has grown in enabling some to evade its barriers. Such migration has been particularly significant in meeting demands for relatively poorly paid and unskilled service work in global cities (Sassen, 1991). In addition, there are growing flows of women as often-illegal international migrants into illegal occupations, such as prostitution, into parts of Europe and the USA. For example, there are an estimated 20,000 illegal female migrants to Greece, mainly from Moldova, Romania, Russia and Ukraine, working as prostitutes there (Hope, 2003). However, they are also significant in manufacturing 'sweatshop' production and in sectors such as agriculture and construction, characterised by seasonal work and fluctuating labour demand.

These forms of irregular and illegal migrant labour therefore help constitute and sustain the least desirable segments of deeply segmented labour markets, characterised by casual, part-time, seasonal, insecure and precarious employment, with pitifully low rates of pay, lack of social security and insurance cover and 'not too many questions asked'. Such people 'provide a pool of casual workers available for virtually any low-grade job at any time and at any place' (King, 1995, 25). While King perhaps overstates the mobility and flexibility of such workers, this does not detract from the more general validity of the point he makes regarding their marginalised and subaltern position in the labour market.

A more recent variant on the theme of attracting international migrant workers to fill gaps in labour markets is 'body shopping', the temporary recruitment of specialised employees on terms and conditions advantageous to the recruiting company. For example, IT companies have recruited Indian consultants to work in the USA and Europe at Indian wage rates plus a subsistence allowance (Pearson and Mitter, 1994). Sometimes this has involved people being brought to the USA or Europe on these conditions to be trained by workers there, who will then be made redundant as jobs are relocated once such workers have received the required training and return home. As well as relocating very large numbers of unskilled jobs in activities such as call centres to low-cost locations such as India, Mexico and Russia, there is an increasing movement of more skilled professional computing, IT and finance workers and managerial staff.

One consequence of the growth of international migrant workers, especially as migrant workers have been drawn from increasingly diverse origins, and of activities such as 'body shopping', has been deepening ethnic labour market segmentation, often intensifying existing tendencies in labour markets already divided on ethnic lines as a result of permanent migrations. Many migrant workers occupy vulnerable and precarious positions, overwhelmingly concentrated in the socially least desirable manual jobs (often because they are 'dirty' and unhealthy as well as poorly paid). For example, in Germany, southern European and North African *gastarbeiter* are heavily concentrated in 'dirty' jobs in the coke works and steelworks, in spot-welding, in foundries and paint shops in auto plants, and in low-level service jobs.[3] This process of occupational segregation strongly resembles the way in which black people in the USA and new immigrants in Australia were also assigned to work in the least desirable and most hazardous jobs (Yates, 1998, 122). In addition, however, immigrant workers are pushed into such jobs because they have largely been excluded from the apprenticeships that are the entry route to skilled jobs in manufacturing in regulated labour markets such as those of Germany. In Italy, a source of many migrant workers for Germany and other north-west European countries, foreign workers are heavily concentrated in agriculture, construction and household services, for example (Brunetta and Ceci, 1998). In the contemporary Californian labour market, Mexican and other Latino immigrants are heavily concentrated in seasonal agricultural jobs such as fruit picking and in the clothing sweatshops of Los Angeles. Chinese immigrants, many of them children working illegally, work in the totally unregulated clothing sweatshops of New York 12 hours a day and for less than $2 per hour in very poor working environments (Harvey, 1996, 287). These divides are often perpetuated by kinship and managerial strategies that encourage the development of ethnic and racial cleavages within workforces.

Such international labour migrations were seen as an acceptable way of averting damaging labour shortages in the core territories, despite (or maybe because of) the way in which they help segment labour markets and divide workers in new ways. However, there are greater reservations about them when they are seen as a mechanism that will lead to the shift of skilled and well-paid jobs to emerging centres of capital accumulation, with much lower wage and production costs. The average weekly wage of an accountant in the UK in 2000 was £630, compared to £30 in India (Turner, 2003). Amid fears of a growing shift of such jobs, there are mounting protests from those losing or in danger of losing their jobs (for example, in the USA: Morrison, 2003). This suggests that there may well be limits to the extent and form of globalised neo-liberal labour markets and international labour migration flows.

5.2.3 Permanent migration in search of work and better living conditions

The key distinguishing characteristic of permanent migration is that the migrants move permanently to new residential as well as work spaces. While a wide variety of motives influence permanent migration, the emphasis here is on those that relate most directly to employment and work, although all such movements of people have economic impacts. Most permanent movements of people involve short-distance migrations (Lucas, 2001). Rural–rural migration remains important, especially in countries characterised by significant employment in agriculture (Lucas, 2001, 328). There are well-established patterns of intra-national age and sex selective migration, from peripheral to core regions, from rural to urban areas, from smaller towns to bigger cities, as people (typically young men but increasingly young women too) have migrated in search of work and better living conditions. Many of these movements have therefore been linked to sectoral economic change, with people moving out of rural agriculture in search of non-agricultural work in urban areas. However, expectations about better jobs have often been unrealised, as rural under-employment has been translated into urban unemployment and/or informal or illegal work in towns and cities (Hudson and Lewis, 1985, 16–18). The growing volume of migrants has been closely linked to processes of suburbanisation, as urban areas expanded to create new residential and consumption spaces, and more recently selective migrations back into the gentrified areas of major inner cities. There have also been growing processes of counter-urbanisation within the more affluent capitalist core countries (Champion, 1989). These involve migration to more rural regions and smaller towns by people who can afford to move to these more desirable residential locations from which they then commute. In addition, those who can afford to do so often choose to retire to more pleasant climatic and environmental surroundings.

While most migration flows are short distance and intra-national, longer-distance international migrations in search of better living or work conditions are also important. Although there had previously been permanent migratory movements on a limited scale, the combination of the expansion of industrial capitalism and of colonialism in the nineteenth century redefined the scale and spatial patterns of such migration. In particular, this involved permanent transcontinental migration from Europe, facilitated by major developments and innovations in transport, with the invention of the steamship. Between 1850 and 1914 some 50 million people migrated from Europe, mainly to north America (70%), south America (12%) and Australia, New Zealand and South Africa (9%). These were voluntary moves, motivated by varying combinations of 'push' (for example, famine, poverty and unemployment) and 'pull' (the promise of new lives and sources of work and wealth) factors. There were important variations between countries and regions in the extent to which they were origin areas for migrants. For example, there was relatively little migration from Belgium, France and the Netherlands, while migration from eastern and southern Europe only began on a large scale from the end of the nineteenth century, with Italy as the dominant source. Within Italy, however, there were significant inter-regional differences, with the Mezzogiorno as the main source of international migrants.

Family links, friendship connections and social networks, rooted in shared spaces of origin, were important in defining the social spaces within which such movements took place, and were often linked to preceding intra-national migration movements in the home country from village to town to city as a prelude to migration abroad.[4] This resulted in processes of chain migration, with emigrants from a particular village or district tending to cluster in the same destination, often performing the same type of work. Kinship links and the desire of emigrants to move into spaces in which they already knew friends and relations who had moved previously reinforced the chain of continuity. Thus major cities in destination countries became ethnic mosaics, with clearly defined residential spaces (Greek, Italian, Polish and so on) and within these spaces there were often more micro-scale concentrations of people from particular villages or regions in the 'old country' (King, 1995, 14–18).

The scale of such movements subsequently decreased and spatio-temporal patterns of movement varied with economic cycles. In addition, people in newly-independent post-colonial states increasingly sought to be allowed to migrate to former imperial core countries while other geopolitical changes led to people wishing to migrate. For example, refugees sought to escape political oppression. However, national states became much more involved in regulating the pace and scale of migration, especially as the economies of the core capitalist countries slowed down, unemployment there rose, and the previously burgeoning demands for unskilled labour subsided. Increasingly, as with temporary migrant workers, national states became more selective in terms of whom they would admit as permanent migrants. Permission to migrate is restricted to people with key skills that are in short supply or to people possessing considerable capital and/or entrepreneurial ability, to whom they are prepared to offer accelerated access to permanent residence and citizenship (Tseng, 2000). There is a growing counter-tendency for those who can afford to do so to migrate from northern

Europe or north America and retire to environmentally more benign climates (such as the Mediterranean or the Caribbean: King et al., 2000). However, this can impose strain upon scarce resources (for example, in health care) in the destination areas.

The migration to Europe of doctors, nurses and medical professionals from parts of Africa, the Caribbean and Asia is well-established and often linked to former colonial ties. The growing tendency for countries in the core of the global economy selectively to use those in the periphery as sources of skilled labour, to meet labour shortages in critical occupations via immigration, has both added to and generalised this tendency. Often, therefore, these migratory movements have led to charges of core countries promoting a neo-colonial 'brain drain' of talented people from peripheral countries, such as Egypt, India or Sri Lanka (all with long-established roles as providers of skilled international migrant labour). In the case of Africa, in 20 countries more than 35% of nationals with a university education are living abroad (Commission on Macroeconomics and Health, 2001, 75–6). This registers an enduring developmental dilemma. While enabling the individuals involved to enhance their lifestyle and acquire more highly remunerated employment, the corollary is that the origin areas lose people with skills that they can ill-afford to lose (for example, in health care and IT), with important developmental implications.

5.3 Moving people, moving knowledge and know-how

Permanent migration was an important conduit for the transfer of knowledges and skills relevant to constructing capitalist economies in new spaces, above all the USA. Migrants to new spaces needed to create new ways of 'getting by' and making a living. As capitalist economies became established, the mobility of people remained an important channel for transferring knowledge between firms and spaces, with temporary migration becoming increasingly significant, linked to innovations in corporate organisation, transport and communications. One consequence is that managers routinely moved between different divisions of firms, taking knowledge with them and acquiring fresh knowledge in the process. Nevertheless, permanent migration remains an important mechanism for knowledge transfer and encouraging entrepreneurialism, not least in some of the most dynamic spaces of the contemporary capitalist economy. For example, Chinese and Indian immigrants have been pivotal in the creation and continuing success of 'Silicon Valley', occupying 25% of all chief executive posts there in the late 1990s (Saxenian, 2000).

In recent years, the relationships between moving people and moving knowledge have increasingly become a focus of attention, particularly given the emphasis upon tacit knowledge and the claims made for it in relationship to economic performance. The movement of people is the only mechanism via which such tacit knowledge can be transmitted between spaces, within and across the boundaries of firms and territories. People who are the repositories of such embodied knowledge therefore become a valuable corporate asset. However, such key employees also become a potential mechanism for knowledge transmission to competitors if they move to another firm. This is especially so in 'knowledge-intensive' activities in which competitive advantage is seen to reside in unique firm-specific knowledge but in which there is intense demand for the key skilled workers in whom such knowledge is embodied. For example, in the 1980s almost 80% of engineers leaving companies in Silicon Valley moved to another Silicon Valley company. Thus the labour market 'serves as a conduit for the rapid dispersal of knowledge and skills among Silicon Valley firms' (Angel, 1989, 108). In such labour markets networked production strategies rely on considerable inter-firm contacts and 'know who' can be critical in recruitment. This has advantages for recipient companies but poses problems for those losing key employees, repositories of sensitive embodied knowledge. Since such tacit knowledge is often seen as *the* key competitive corporate and territorial asset, companies are seeking ways of codifying knowledge to help insulate them from the effects of losing key staff. This prevents expensive knowledge central to the core competencies of a company that differentiate it from its rivals being lost whenever employees leave or die.

More generally, the movement of key personnel within and between companies and/or public sector organisations is assuming greater significance as a mechanism of knowledge transfer and learning (Mahroum, 1999). These form a 'new group of executive nomads' (King, 1995, 24). There are three identifiable groups that play key roles in this process of nomadism. First, there are global bureaucrats, executives of the World Bank, the International Monetary Fund and the World Trade Organisation, who have a critical involvement in specifying the framework conditions of the internationalisation of capital. For example, such people are involved in imposing structural reform programmes on peripheral states, often as a condition for eligibility for trade and aid packages. Linked to these, there are specialists working on contract as part of such packages. Secondly, there are transnational corporation executives, managers, engineers and technicians, mobile workers who transfer knowledge and standards and seek to cascade current notions of discursive and material 'best practices' across locations within corporate structures, either for direct use or for transmitting relevant knowledge to 'local' workers via training, if need be suitably modified in response to corporate cultures and/or local specificities. For example, Hardy (2002, 174–84) describes how ABB (Asea Brown Boveri) and Volvo adapted very different approaches to establishing new management cultures in the same part of Poland in the 1990s: Volvo adapted a gradualist and pragmatic route, while ABB's approach was more akin to 'shock therapy'. The movements of key executives may transfer knowledge of particular practices across the boundaries of firms within collaborating networks or strategic alliances, between customers and suppliers or between the producers and users of innovations. Finally, there are the professionals, particularly consultants – the unacknowledged legislators who produce and disseminate current versions of 'best practice' within the 'new circuits of cultural capital' (Thrift, 2001). As well as management consultants and global head-hunting firms, private sector temporary employment agencies are also becoming more involved in organising movements of highly qualified and skilled professionals, either hired on permanent contracts or as consultants to particular companies to carry out a specified task or project, in both cases mobile across both corporate and national boundaries.

Consequently, there has been a great increase in the volume of business travel and this 'constant quartering of the globe by executive travellers ... seems to be the result of attempts to engineer knowledge creation' (Thrift, 2002, 221–2). This engineering process is oriented towards settings which privilege face-to-face communication, 'the richest multi-channel medium because it enables use of all the senses, is interactive and immediate' (Leonard and Swap, 1999, 160, cited in Thrift, 2002). Rather than improvements in ICT leading to the reduction of such creative interactions via face-to-face meetings, they in fact enable them. This is because they allow managers to keep in touch with their home base: 'rather than the home base being a means for the control of the periphery – the home base comes visiting. Control, or more accurately modulation, is executed through the circulation itself.' Moreover, the 'new jet setters who circle the earth not only via jet but also by cellular phone and e-mail, serve not only to promote the international circulation of financial products and industrial goods, they also serve to promote the circulation of liberalising ideas' (Schmidt, 2002, 39).

5.4 Flows of economic migrants, reverse flows of remittances and return migration

A corollary of flows of people becoming and performing as international labour migrants in search of work is a regular reverse flow of money back to their origin areas. This typically helps sustain the consumption levels and lifestyles of their families there, and is used to construct new houses to which long-term migrants eventually retire on return. It is less common for return migrants to use accumulated remittances to invest in new economic activities (other than purchasing land) as a way of providing new sources of (self)employment and streams of income in their areas of origin, although younger returnees to urban areas are more likely to do so (Hudson and Lewis, 1985, 23–5; Sampson, 2003, 270–1). Even so, most such investment is in service activities such as bars, cafés and taxis, tied into local circuits of exchange and consumption.

However, remittances nonetheless often have a macro-economic significance. In recognition of this, some national states require the remittance of earnings abroad. For example, the Philippines Overseas Employment Agency requires migrant workers to remit a minimum of 80% of their basic earnings to a Philippines bank account (Sampson, 2003, 258). Globally, annual migrant remittances have been estimated by the IMF to have increased from $599 million in 1970 to $61 billion in 1998 (Williams et al., 2004). More recently, the World Bank estimated annual remittances at $80 billion (International Organisation for Migration, 2003), recognising that the actual volume of flow may be two or three times this sum. However, even the lowest-level estimate exceeds the total for official developmental aid. These flows of money are mediated via banks, linking spaces of origin and destination, of work and home, but migrant workers also remit money in a variety of other, less traceable forms, which contributes to the difficulties of recording all flows. Remitting money in these varied ways is linked to the experience of spaces of work as alien, as a space to which migrants feel only an instrumental, one-dimensional attachment to earn money, not a multidimensional attachment to the meaningful space of home. The relationships between spaces of work and home may become more ambiguous with longer-term migrants, however, especially as they change, perhaps constructing more hybrid identities, and so develop a more ambiguous stance to the space called 'home' or, indeed, become 'homeless'. Equally, return migrants return to spaces that have changed in their absence, just as they have changed.

5.5   Consumption, exchange and flows of people

For many people, especially in rural spaces in the peripheries of the global economy, everyday life is conducted within restricted spaces, limited by the distances that they can cover on foot. Everyday life remains, in that sense, highly localised, even if images from and of life in other parts of the world regularly arrive via television and other media as people engage in 'weightless travel' (Urry, 1999). For the majority who live in urban areas, however, the flows of everyday life are typically more complicated, more technologically mediated, as people have access to a range of modes of private and public transport. However, the dominant mode is the automobile, reflecting the pervasive power of the 'road lobby' (Hamer, 1974) and the centrality of automobility to capital accumulation. The demands of automobility structure the land use patterns of cities, on occasion to an enormous degree: for example, 50% of the land in Los Angeles is devoted to car-only environments, as is 25% of that in London. This reflects the dominant modernist forms of land use/transportation planning in shaping such spaces, which are also characterised by the spatial separation of different functional uses. As well as the daily routine of journey-to-work, people need to make regular journeys to shop, to leisure facilities of various sorts, to schools and other educational establishments, to doctors' surgeries and hospitals and so on. In all cases, however, these flows both link and create spaces of various types. These trips are typically made with different frequencies and in different directions and distances from residential spaces, indicative of the extent to which literally 'spaced out' urban environments have developed, deliberately or inadvertently, to maximise the distances between required services and facilities. People may strive to combine different activities into multi-purpose trips, but this is often difficult, especially if it involves using a variety of modes of transport, especially public transport. The attraction of the automobile as a private mode of transport (whatever its environmental and social costs) is that it allows the flexibility to juggle time/space and link together disparate functions in single, individually customised, multi-purpose trips. The motives for different flows reflect the fact that people perform multiple roles in the economy and assume associated identities. Different roles and identities dominate in particular flows and time/spaces, as people perform as workers, purchasers, consumers, family members and so on. The automobile thus permits these multiple socialities of community, family life and leisure, as well as those of the workplace, shopping mall, doctor's surgery, children's schools and so on, which are interwoven through complex jugglings of space/time that car journeys allow. Much of what many people now regard as 'social life' – not least, touring around and seeing a variety of environments, which itself constitutes a source of pleasure and recreation for many – would be impossible without the flexibility offered by the automobile and its availability 24/7 (Urry, 1999, 8).

As well as these everyday, frequent and generally short-distance flows, people also are involved in activities that require less frequent and often longer-distance flows, with specific institutional and corporate forms developed to enable them. International flows of tourists exemplify this, as tourism has expanded massively in scale,[5] accounting for around 10% of both global employment and GDP. Certain organisational innovations and material investments have transformed the nature of travel in ways that have been highly socialised and at the same time inscribed it into circuits of capitalist production. The emergence of the modern travel agent, with the formation of Thomas Cook in the 1840s, was particularly important in this regard. For Cook was responsible for a number of product innovations which transformed travel from something that was individually arranged, risky and uncertain into one of the most organised and rationalised of human activities based upon considerable professional expertise (Lash and Urry, 1994, 254–63). Subsequently, Cook was centrally involved in the post-1945 growth of package holidays. The emergence of the package holiday in the nineteenth century depended upon the development of the railways and the growth of city-centre hotels adjacent to railway stations. The development of mass tourism package holidays in the second half of the twentieth century depended upon innovations in air transport (although for many the automobile provided the means of transport to and from airports) and the developmental aspirations of national states on the peripheries of the global economy. The net result was that mass tourism penetrated new spaces for the first time (Young, 1973). Often this transformed them irrevocably – perhaps best (or worst) exemplified by the high-rise concrete constructions of the littoral of the Costa del Sol in Spain. As such development typically took place in previously sparsely populated areas, flows of people as tourists in turn served to generate flows of people of migrant workers, typically on a seasonal basis, and with deleterious consequences on their areas of origin in activities such as agriculture.

Flows of people as travellers and tourists reflect and require sequences of exchange relations (Lash and Urry, 1994, 270–1) and so are linked to circuits of capital and money. First, this mobility of people involves the exchange of money for temporary rights to occupy a mobile property, such as an airline seat or a cabin. Secondly, it involves the exchange of money for temporary rights to occupy a living space – a rented villa, a hotel room, or a restaurant table. Thirdly, it involves the exchange of money for sensual property (such as smells, views or sounds) as a corollary of the acquisition of temporary rights of possession.[6] A combination of organisational, technical and travel innovations has led to growth in both the volume and spatial reach of tourist flows. This has involved extending the scale of flows from the intra-national to international and intercontinental. It has also led to more complex spatio-temporal patterns, drawing in a greater variety of locations to become destinations and offering a greater range of experiences between and within these tourist spaces. Furthermore, the range of spaces of tourist origins has also become more varied, as high-income social strata, keen to perform the role of international tourists, have emerged in peripheral countries, while the increasing growth in second and third annual holidays has also led to a more complex, and in some ways less temporally peaked, pattern of flows (Shaw and Williams, 1994). This increasing variety of both destinations and origins is linked to growing market segmentation, with particular types of tourist flowing to particular types of tourist space, possibly at particular times of the year.

The expansion of mass tourism, and more generalised forms of customised tourism, has provoked a reaction, a counter-tendency to post-tourism as people seek again to become travellers but with many less risks than in the pre-Cook era. More individualised flows, detached from the masses, the mob and the throng, have been facilitated by the growth in ICT and the possibilities of arranging flights and booking accommodation directly via the world wide web. In addition, the explosion in available information via a range of electronic media (not least TV travel programmes), the spectacular growth of simulated travel and the extraordinary proliferation and circulation of images and signs, has opened up information about new potential spaces for post-tourists and travellers. It is important to remember, however, that such 'post-tourism' remains limited to specific social groups, defined on the basis of age and/or available income, with those not yet engaged in the mainstream labour market or the relatively affluent and retired from the labour market most able to engage in such forms of tourism. For example, some young adults perform 'backpacker tourism', typically in spaces on the periphery of the global economy, although most take their holidays as part of organised packages (Hampton, 1998).

5.6 Summary and conclusions

The performance of the economy both requires and results from flows of people in a variety of capacities and roles – as consumers, purchasers, producers, workers and so on. Since activities of production, exchange and sale, and, to a considerable extent, consumption, are collectively performed, people must move between the varied sites in which these activities and practices take place. Moreover, they must move between them in the 'correct' sequence. People must be in the appropriate time/space to ensure the required sequence of activities that make the economy possible. These movements entail enormously complicated patterns of flows, of varying temporal durations and spatial reaches, requiring considerable investment in the transport infrastructure, in the means of transport, and in the organisational capacity to ensure that these flows continue smoothly. Mass movements of people thus necessarily require a sophisticated material infrastructure, especially given the 'spaced out' character and complex spatialities of economies.

Equally, people as active and knowledgeable subjects are necessarily central to the constitution of the economy. They have complex motivations and variable knowledge of spaces of economies and the possibilities that these might offer to them (and often their families and friends) in terms of employment, wages and living environments. They also have variable capacities (in terms of money, power and influence) to realise their ambitions and aspirations because of their position in social structures of authority and power. One expression of this is the complex patterns of movements in which people engage, often varying over a life-time, as they seek out opportunities for employment, more desirable environments in which to live, exciting spaces to explore on holiday and so on. While a few people are relatively unconstrained in such movements, most people are limited to varying extents in their choices. For many people such constraints remain severe in the extreme.

Notes

1 Norman Tebbit, one-time Thatcherite Secretary of State for Employment, advised the unemployed to 'get on their bikes' and search for work elsewhere.
2 Even so, growing numbers of frequent international travellers may magnify the transmission of infectious diseases and add to problems of public health (Commission on the Macroeconomics of Health, 2001, 76).
3 Following German re-unification, there were calls for the repatriation of migrant workers to create employment opportunities for residents of the former East Germany.
4 One consequence of the combined effects of migratory movements is that over 50% of the world's population now live in 350 mega-cities with a population in excess of one million people (Soja, 2003).
5 Defining a 'tourist' remains a practical problem. Currently there are 600 million international passenger arrivals annually, a mix of business travellers and holiday-makers.
6 Thus tourist flows involve the consumption of tourism spaces (see Chapter 9).

6　Spaces of Regulation and Governance

6.1　Introduction

The spaces in which 'the economy' is made possible must be politically and
socially (re)produced. The complex and potentially unstable ensemble of practices
and processes denoted by 'the economy' is made reproducible via territorially
constituted and multi-scalar processes of regulation and governance, centrally
involving but not limited to the activities of the state. These multi-scalar spaces
of governance provide the political framework within which spaces of production,
sale and consumption can be constituted, governed and regulated and through
which flows (of capital, information, money, people) between and within these
spaces can be regulated. The economy is thus constituted in and through a series
of scaled territories – cities, regions, national territories – constructed *as* closed-
off, bounded but nonetheless permeable spaces, allowing regulation of cross-
boundary flows as well as of economic relations and practices within these
boundaries. While these boundaries are not permanent, fixing regulation at particular spatial scales provides a degree of stability to the formal institutions of
the state. This is a necessary territoriality, with social relations sedimented into
specific spatial forms. In this way, spaces of governance become the (in part constitutive) basis for territorially specific forms of capitalism, linking consumption,
exchange and production in particular ways. Such spaces are representational,
making visible both objects and subjects of governance. Their legitimacy is to
some extent linked to collective perceptions of them as encompassing territorially defined 'imagined communities' (Anderson, 1982) and to a recognition that
particular conceptions of governance are hegemonic or, alternatively, are maintained via coercion within their boundaries.

Two further points can be made by way of introduction. First, state actions
and practices are strategically selective (Jessop, 1990). There are important relationships between structural parameters, institutional forms and agents in shaping selection mechanisms and strategies which impose filters on the content of
policy and upon who is included in and excluded from debates about policy and
the policy agenda. Forms of selectivity are linked to modes of governmentality,
ways of conceptualising both the objects and processes of governance and regulation. In particular, state actions are spatially selective, with a tendency to privilege

certain spaces within accumulation strategies (Jones, 1997, 832). This is partly because of political pressures to be seen to be concerned with territorial equity but there are also structural limits, inherent to the accumulation process and in the relationship of the state to capital, as to its scope to exercise selectivity in this way. The continuing dependency of the state upon the accumulation process establishes limits to its capacity to act and to manage crises autonomously (Offe, 1975, 144). In order to try to cope with this, the state deploys a socially specific 'sorting process', influenced by the interests of dominant classes and historically-contingent specific state functions, including some social groups in processes of policy formation while excluding others. However, while strategically selective, it is important that state policy is represented and popularly accepted as a hegemonic project, one that is in the general interests of all citizens within a given territorially defined community. Hegemonic status thus helps disguise the socio-spatial selectivity of state policy agendas and actions. Strategic selectivity includes the ways in which the organisation of the state differentially privileges the access of some forces of representation to formulate specific policies and secure their support within the state apparatus and then to implement these policies effectively (Jessop, 1990).

Secondly, conflicts and tensions may impede regulation and governance. There are two sorts of reason as to why this may be the case (Painter and Goodwin, 1995, 342). First, such tensions may be partly the (un)intended products of actions undertaken by individuals and institutions, a consequence of the emergent properties of practices. Secondly, there may be counter-tendencies operating to contest and disrupt the reproduction of capitalist order. For example, people may seek to circumvent or break rules rather than comply with them. Emergent properties may produce disruptive effects, interfering in the smooth operation of regulatory processes. Because governance and regulation are multi-scalar processes, such unintended effects can be expressed at various scales. Effects may be as intended at one scale but unintended at another. There is no guarantee of scalar compatibility in this regard. Indeed, a particular scale of regulation may itself be an unintended product, as, for example, locally based resistance to national regulatory action creates 'local' spaces of resistance that themselves then become objects and subjects of regulation.

6.2 The national as the privileged space of regulation and governance

National states are involved in myriad ways with the regulation and governance of national economies, affecting production, exchange and trade, and consumption both within and between states (R. Hudson, 2001, 76–92). As such, the national is a critical regulatory space and scale. There are many ways in which state policies relate to the economy, including:

- providing material infrastructure to enable production, transport, exchange, and consumption of commodities;
- providing ICT infrastructure to link buyers and suppliers;
- regulating product markets, product quality and prices;
- regulating competition between companies and between capital and labour;
- encouraging the creation of new products and production processes by R&D and innovation policies;
- influencing international trade patterns via trade policies and policies relating to foreign direct investment;
- shaping labour markets via policies on trades unions, education and training, and migration;
- regulating the labour process via legislation on issues such as health and safety at work;
- shaping income distribution via prices and/or incomes and taxation policies;
- influencing the location of activities within the national territory via financial incentives to locate in some places and legislation to prevent location in others;
- seeking to reduce pollution, set environmental standards, and encourage re-cycling via environmental legislation; and
- selectively replacing markets as resource allocation mechanisms and partially decommodifying the production of some goods and services.

The list, though long, is nevertheless indicative rather than definitive.

This practical involvement of the national as a pivotal regulatory space and of the national state in regulating national economies is well known. The national state has been discursively constructed as having absolute authority and mastery over its national territory, recognised as sovereign by other national states (although this modernist conception of the national space has been regularly violated by (neo)imperialist adventures) and, as such, able to define and manage a national economy. In the 1970s, the state derivation debate sought theoretically to specify the characteristics of *the* capitalist state within the abstract space of capitalist social relations (Holloway and Picciotto, 1978). However, it quickly evolved to recognise that the prime regulatory spaces of capitalism were constituted as the contiguous and bounded territories of national states, acknowledging the geopolitics of capitalist development and *the* key regulatory role of the national state in this.[1] National state, economy and society are mutually constitutive, with shifting and permeable boundaries between them. For example, the retreat of the state from pension commitments over much of the European Union is encouraging and coercing people to provide their pensions via the market. This is creating space for pension fund managers to increase the role of private capital in pension provision, and so the significance of pension fund managers in determining circuits of capital.

As well as seeking to regulate the economy within their national territory, national states also have connections to other such states. They must manage links with other national states in international regulatory bodies (such as the

IMF and WTO), with embryonic supra-national states (such as the European Union) and with extra-state organisations (such as multinational companies). Location in the international state system has important constraining effects on room for manoeuvre in policy formation and implementation. The scope and content of state policies is open to multiple sources of influence, with many proximate causes mutually determining trajectories of state activity. States have to balance two sorts of demands in formulating and implementing their economic policies. First, they must ensure the smooth accumulation of capital in their national territory. Secondly, they must be seen to act legitimately, to preserve both the authority of the state and the hegemony of capitalist social relations. Seeking to satisfy these contradictory demands creates problems for the state in carrying out its own activities. While exercising power with the objective 'in the last instance' of reproducing capitalist social relations, the question of *how* the state seeks to do so, *how* it walks the tightrope between the competing demands of accumulation and legitimation, remains open.

The constitution of capitalist states as national states formed the starting point for various regulationist accounts of how states sought to perform this balancing act.[2] Regulationist approaches seek to navigate a course between 'agentless structures and structureless agents' (Jenson, 1990) and envisage a definite relationship between the national economic growth model (or regime of accumulation) and the ways in which this is regulated (mode of social regulation). Regimes of accumulation are relatively stable aggregate relationships between production and consumption; modes of regulation are the regulatory mechanisms within state and civil society, the body of beliefs, habits, laws and norms consistent with and supportive of a regime of accumulation. The national state balances increases in productivity and consumption within the national territory via public expenditure, taxation and income redistribution policies. Regulationists emphasise the correspondence within a given mode of development between regimes of accumulation and modes of regulation necessary for social systemic stability. As such, a mode of regulation is a set of mediations that contains the accumulation process within limits compatible with social cohesion in a given territorial formation (Aglietta, 1999, 44). However, establishing a mode of regulation also involves a 'representational struggle' as to the most appropriate way of coupling consumption and production, thus emphasising its political character (MacLeod and Jones, 1998).

Such couplings provide a spatially specific temporary resolution to the inherent structural contradictions of the value form within capitalist social relations, encompassing both commodity and non-commodity elements needed to underpin growth and accumulation. While the state generally plays a key role in securing a given mode of social regulation, this also necessarily involves the participation of capital and a variety of groups within civil society. Political practices, social norms and cultural forms contingently combine in complex ways to enable the dynamic and unstable capitalist system to function, at least for a period, in a relatively coherent and stable fashion.

Consequently, modes of regulation do not simply exist, pre-formed and awaiting discovery by national states: national states are not 'the prisoners of a fixed genetic code' (Weiss, 1997, 18). They can search for and adopt varied forms and modes of regulation. Indeed, such experimentation with regulatory projects is likely to be an ongoing process, particularly intense during periods of crisis as the state grapples with 'discovering' or inventing an appropriate mode of regulation that enables competing demands to be made temporarily compatible. Compatibility needs to be achieved on two levels. First, in terms of the different 'political shells' within which state structures are situated, broadly divisible into democratic and non-democratic forms, depending upon the extent to which state authority and power is secured by consensus and persuasion rather than coercion and the exercise of the state's legitimate monopoly of physical force and violence within the national territory. Jessop (1982, 233) refers to these as, respectively, 'normal' and 'exceptional' regimes. Secondly, within any particular 'political shell', compatible combinations of growth models, forms of economic organisation and modes of regulation must be discovered. There are, for example, national variations around the canonical generic Fordist regime of accumulation (Lash and Urry, 1987). Even with the added nuance of national variations, however, formulations of a dichotomous progression from Fordism to post-Fordism are deeply problematic (Hudson, 1989; Sayer, 1989). Nevertheless, Lash and Urry (1994, 63 ff.) go further and contentiously link a transition to post-Fordism to the rise of reflexive accumulation, identifying three 'ideal types' of post-Fordist reflexive accumulation, associated with different national regulatory modes based on Japan, German-speaking Europe, and the USA and UK.

In summary, modes of social regulation are formed within determinate limits through indeterminate political and social struggles. The establishment of a stable coupling between an accumulation system and a mode of regulation is contingent, a chance discovery in a specific time/space context. Some combinations of regime of accumulation and mode of regulation are feasible, while others are infeasible. Each national state seeks to support a particular economic growth model and sustain a particular set of social and political bargains and compromises within its sovereign space of regulation to underpin this. Consequently, national spaces can exhibit quite distinctive couplings of accumulation model and mode of regulation in the same global political-economic environment. The contrast between the UK and USA, West Germany and Sweden, and Japan in the 1980s exemplifies this. Discovering and securing a feasible mode is far from straightforward, however. Reflecting the complex and multidimensional character of capitalist societies, and the variety of interests represented within them, the state is confronted with a demanding policy agenda. Internally, it seeks to balance competing claims over accumulation strategies and economic growth trajectories, and deal with issues of equity and social justice in the distribution of the benefits and costs of growth. As capitalist societies have become more complex, the balance of issues on the agenda that the state sets for itself or has set for it by other social forces has altered. For example, ecological issues are now much more prominent. There have also been significant changes in the ways in which national states approach their roles of crisis avoidance and management in order to try to guarantee relatively long periods of economic prosperity and social stability.

6.3 Crisis tendencies and state regulation: (re)defining the boundaries between economy, society and state

Regulationist accounts are curiously silent about transitions between regimes of accumulation 'beyond describing them as intervals of crisis' (O'Neill, 1997, 297) and about the role of the state in moving societies from one period of economic stability to another. They skate over the processes of transition between modes of regulation and of crisis management and resolution that this implies. Crises are apparently resolved without difficulty – a new stable combination of regime of accumulation and mode of regulation mysteriously emerges fully fledged and functioning. Others, in contrast, see such stability as rare in a world uneasily held together by contradictory regulatory practices. Conceptualising regulation in terms of shifts between periods of stability is thus misconceived (Painter and Goodwin, 1995). Moreover, in practice, seeking to resolve crises is difficult and problematic and the state struggles to resolve competing claims, articulated via a multiplicity of sources of influence that attempt to shape the content and scope of its activities and policies. In such moments of struggle, paradoxes and inconsistencies become more evident, while experimentation intensifies until a new mode of regulation sediments into place.

State activity becomes unavoidably problematic and crisis prone in part because of the involvement of the state in the affairs of other institutions and associations, such as political parties, trade unions and corporations, and in the processes through which economic and social interests are represented to government. Conflict and tension is chronic as groups with different and competitive logics seek to influence the course of, or be included in the domain of, state actions. The consequences of dealing with social turbulence and political resistance (for example, to rising taxation levels or in response to fears of rising inflation) are continuously, chronically and routinely internalised within the state apparatus. This is part and parcel of ongoing attempts by the state to manage and distribute resources in ways that satisfy prevailing notions of social justice and moral rectitude as well as the achievement of economic growth (O'Neill, 1997, 296). The state is neither an arbiter, nor a regulator nor an uncritical supporter of capitalism but it is 'enmeshed' – and *unavoidably* so – in its contradictions (Held, 1989, 71, emphasis added). For the social processes necessary for the reproduction of capitalist economies 'are regulated and sustained by *permanent* political intervention' (Offe, 1976, 413, emphasis in original).

The transition from a liberal to an interventionist mode of state operation (Habermas, 1976) sharply emphasises the problematic character of state involvement.

A liberal mode of state involvement sets the parameters within which the law of value operates in a particular state territory, facilitating the variably (in)visible hand of the market in allocating resources. An interventionist mode of state involvement unambiguously reveals the visible hand of the state, supplementing or replacing market mechanisms; for example, via nationalising key sectors of industry or assuming responsibility for educational and health care provision. The proximate motives for this major qualitative extension of state activities vary. In the final analysis, however, it is because such activities are no longer (sufficiently) profitable to attract private capital but they are seen as providing goods and services that, for a variety of reasons, 'must' be provided, realigning the boundary between private and public sector provision (Hudson, 1986). Defined in this way, all national states in the advanced capitalist world are now, to varying degrees, interventionist. Nevertheless, there are national variations in the extent and forms of involvement, and in the political projects and strategies that inform this.

However, provision of vital goods and services does not become economically non-problematic simply because it is taken over by the state. It results in the displacement of endemic crisis tendencies from the economy, internalising them within the state and its mode of operation. Consequently, the underlying economic crisis re-emerges within the operations of the state or within civil society. Rationality crises arise because of competing pressures seeking to shape state action: for example, those of efficiency versus equity. As a result, there often are unintended (and overtly spatial) effects as well as – or instead of – intended outcomes, in part because regulatory processes interact contingently with existing patterns of uneven development and historically prior uses of space. Legitimation crises may emerge, or at least be threatened, because this chronic gap between actual and intended outcomes generates tensions, which may challenge the territorial integrity and/or the authority of the national state and the dominant pattern of social relationships represented through it. This leads the state to seek strategies to reassert the legitimacy of its actions and maintain the unity of its territory. Fiscal crises arise because decommodifying the production of goods and services under conditions of parliamentary democratic representation and bureaucratic policy making constantly acts to increase fiscal pressures on the state because it seeks to subvert the logic of the market (Jessop et al., 1988, 160). Moreover, O'Neill (1997, 299) argues that tendencies towards fiscal crisis are endemic precisely because the state accepts responsibilities for *ex post* income distribution – and no democratic state can be wholly indifferent to this, for reasons of political legitimacy and electoral calculation. Consequently, the state will 'always' experience fiscal crisis during downturns in the economic cycle, when there are increased distributional demands and falling revenues. Moreover, such crisis tendencies will be even more marked during the downswings of long waves and the troughs before the turning point to the upswing of a new wave. As a result, the simultaneous successful performance of state functions is impossible for any length of time and so the state has little choice but to confront conditions of fiscal crisis directly.

Resolving fiscal crises is perhaps the most difficult operation in crisis avoidance and management. Reining in the extent and form of state activity and reducing the resources needed to sustain it runs the risk of triggering a legitimation crisis. However, the immediate political dangers of such a crisis are less serious than those of a generalised crisis of accumulation, although a prolonged legitimation crisis would almost certainly undermine accumulation. Not least, successful accumulation is critical for funding state activities, setting a limit to public expenditure, as well as acting as a selection mechanism in influencing the distribution of that expenditure. Consequently, the boundaries between public and private sectors may be redrawn to reduce state involvement in economic and social life. This recommodifies the production of goods and provision of services such as education and health, while introducing pseudo-profitability efficiency criteria into the remaining public sector industries and services and displaces other sorts of service provision back into the household. Thus 'the four centrally important strategies of restructuring of state services will be intensification, commodification, concentration and domestication' (Lash and Urry, 1994, 209). Such moments of boundary redefinition between private and public sectors, between state, economy and civil society, may herald the transition to a new regime of accumulation and mode of regulation and mark significant turning points in developmental trajectories. Or they may not. Whether this actually has been the case can only be known *ex post facto*. The general implication is nevertheless clear: crises will appear in different forms as the content, form and style of state policies and politics varies while policies and politics will be altered in response to crises and also less dramatic changes in the wider political-economic environment.

However, the state comprises a plurality of institutions, organisations and agencies (departments of central government, local government, other state agencies and so on) and the ways in which they relate to one another and function are constituted politically. There may be intra-state conflict in deciding the policy agenda and the scope and content of policies in response to specific problems of governance and regulation that confront particular state agencies and organisations. As a result, the state seeks to restructure *itself* in response to perceptions about the changing external environment and the problems of dealing with qualitatively different and complex problems simultaneously: 'every time a state deals with a problem in its environment, it deals with a problem of itself, that is, its internal mode of operation' (Offe, 1975, 135). The state seeks to reconcile the contradictions inherent to its involvement in economy and society by altering its internal structures and modes of operation and procedures for 'problem solving'. Moreover, the state is both an arena for conflict between classes and groups in civil society and an organisational structure through which they can pursue their own interests and which shapes the ways in which they do so. Different classes, class fractions and other interest groups have variable access to the organisations of the state and differential opportunities to influence the state policy agenda and priorities.

A national state necessarily pursues qualitatively different objectives *simultaneously* in the face of a barrage of externally originating competing and maybe incompatible demands and priorities to which it must respond. Simultaneously, it generates competing demands internally, a consequence of its constitution as internally differentiated and heterogeneous. It operates in an uncertain, contradictory and conflict-ridden environment, one prone to create situations in which it will be unable to satisfy all the demands made upon it and all the expectations that it may raise as to its capacity to solve problems. Consequently, the state may become 'overloaded' (Cerny, 1990), increasingly unable to deal with the demands made upon it because of enhanced flows of money capital, commodities and information in an increasingly open and global economy that erode the efficacy of its 'traditional' tools of national economic management. The underlying class relations that structure state form and shape state policy priorities set limits to but do not determine, logically or historically (Hirsch, 1978, 107), the relations between organisational structure and objectives and set limits to the state's room for manoeuvre in seeking more efficacious policy approaches.

In summary, the national state is engaged in a project of permanent crisis management, since it cannot abolish the crisis tendencies inherent within the capitalist mode of production but only internalise them, at least in part, and contain them for a time. The ultimate goal of the state is the pursuit of stability, negating the potential political implications of economic crises and the contradictory social relations of capital. The boundaries and content of state actions are products of political struggles and there have been important changes in the way states have approached regulation. The change from a liberal to an interventionist stance represented the discovery of a new feasible mode of regulation, associated with a switch from an extensive to intensive regime of accumulation. However described, the net result was growing state involvement in more spheres of economic and social life in much of the advanced capitalist world; in short, in varied 'corporatist' arrangements. This further blurred the boundaries between economy, state and civil society there. As the extent of state activities grew, the state *itself* increasingly became the locus of conflict, a participant in conflict (often waged between its own departments and organisations).

However, as the growing prominence of neo-liberal state forms in the last two decades of the twentieth century sharply demonstrated, extending state activity was neither inevitable nor irreversible as the state pulled back and allowed more space for market allocation. This in turn has been associated with a shift in emphasis in many national states from direct involvement in producing goods and providing services to an enabling and facilitating role. Such states seek to support initiatives from within civil society to provide new sources of employment and service provision in interstices vacated by the state and which the private sector finds unattractive. The growth and subsequent retreat of state activities in the advanced capitalist states is indicative of the limits to national states' capacities to manage, let alone resolve, crises. The emergence of fiscal crises threatened accumulation in national territories. Conversely, there are lower limits below which state expenditure and involvement cannot fall (within a parliamentary democratic state at least) if the legitimacy of the state is to be maintained and, more prosaically but no less importantly, if political parties are to win elections. Recognising these limits, national states sought their own 'spatial fixes' by redefining geometries of regulation and governance, encouraging and directly supporting new supra-national and sub-national spaces.

6.4 'Reorganising' national states: scalar shifts and the avoidance, displacement or resolution of crises?

Regulationist approaches initially tended to give methodological and ontological primacy to the national level. This was assumed to be the sole – or at least pre-eminent – site of state regulation. This claim has increasingly been seen to be unsustainable, theoretically and empirically. Not least, as regulationist approaches evolved there was recognition that they were expressive of a particular discursive construction of the state, its relations to economy and civil society, and its spatiality. As such, they had their own limits. There is no necessary reason for regulation to be primarily or predominantly a national scale process. Spatial scales are constituted through political and social struggles and particular representational practices (Jones, 1997). Recognising this spatio-temporal specificity and contingency, Hay and Jessop (1995, 305) emphasise the 'constantly evolving spatial forms of accumulation and regulation', the dynamic of spaces of governance and regulation since these can never be more than a temporary fixing of the contradictory social relations of the economy in a particular spatial form.

As national growth models and national states policies slipped into crisis, undermined by new and growing forms of globalisation, the silences of regulation and state theories about the re-allocation of state powers upwards, downwards and outwards became increasingly problematic. Some linked this rescaling of governance and regulation to the alleged emergence of reflexive accumulation, made possible by developments in ICTs and tinged by a problematic technological determinism (Lash and Urry, 1994). While acknowledging the continuing significance of the national, this must be located in the context of a visibly shifting architecture of state power and mechanisms of regulation and governance. Recognising this, Peck (1994, 155, emphasis added) argues that social structures of accumulation and regulation are relatively autonomous, yet bound together in a necessary relation. Consequently, their causal powers will be realised in different ways, dependent upon contingent circumstance so that 'the nature of the regulation–accumulation relationship *is qualitatively different at each geographical level*'. This suggests a distinctive multi-level geography to the ways in which accumulation and regulation are linked, involving qualitatively different and, to a degree, scale-dependent types of relationship.

The 'reorganisation' of the state encompasses a triple process of de-nationalisation (hollowing out[3]), de-statization of the political system and the

FIGURE 6.1 Creating a new supra-national regulatory space: re-drawing the map of Europe in 1992

internationalisation of policy regimes (Jessop, 1997). The extent to which regulation is carried out at sub-national and supra-national scales is related to pressures on the national state form 'from above' and 'from below'. Pressures 'from above' resulted from growing economic globalisation, tendencies towards the transnationalisation of political organisation, and the enhanced significance of regulatory bodies such as the IMF or WTO. The erosion of the efficacy of national monies and the national as a space of monetary regulation accelerated from the early 1970s with the collapse of the Bretton Woods arrangements and an increasing tendency to create new global financial spaces, monies and regulatory systems. A combination of 'securitization, deregulation and electronification' (Lash and Urry, 1994, 18–22) facilitated radical transformation of markets for capital, credit and money, with some product markets operating globally and on the basis of 24-hour-a-day trading. This further exposed national economies to the combined regulatory pressures of markets as steering mechanisms and global monies as disciplining technologies. The undermining of the national economy as an object of state management, 'notably through the internationalisation of trade, investment and finance' (Jessop, 2000, 4)[4] was a major factor contributing to the 'crisis of Fordism'. The emergence of floating exchange rates, digital money and electronic transfer systems seriously compromised the capacities of central banks and national governments to control their own currencies (Warf and Purcell, 2001). The creation of the Euro by eleven member states of the European Union involved a voluntary denationalisation and Europeanisation of their currencies, partly with the intention of this providing a stronger currency globally via rescaling the space of monetary governance. Furthermore, there are new forms of 'denationalising' monies and credit systems, constructed in particular ways and controlled from specific local spaces, such as global cities but also offshore regimes (Leyshon, 2000).

There has certainly been a resultant diminution in the capacity of many national states to control monetary and fiscal policy. Equally, however, the national remains critical in relation to money as a pivotal source of social power. For this depends upon it being a *privileged* means to control access to wealth. While money as a representation of value can and must circulate freely, as social power it depends upon a territorial configuration and socio-political system (in short, a national state) that renders that particular form of social power hegemonic rather than occasional and dispersed (Harvey, 1996, 235). The relative denationalisation of the world economy is thus a contradictory process, with the creation of global markets actively authored by national states (Panitch, 1996), which continue to form key sites of global market regulation. Thus 'it is not a question of whether capital's internationalisation *results* in the decline of the state, but rather how the state continues to *participate* in *capital's* internationalisation in order to reproduce itself (Yeung, 1998, 296). Moreover, the weakening position of the national state in fiscal and monetary policy also reflects the growing capacity and power of private sector companies to create credit and a global credit system that is largely private and subject predominantly to market regulation (Altvater, 1990, 23). These new global forms of credit in a debt and paper economy are beyond the control of national states and exert great structural power over them (Leyshon and Thrift, 1992), reducing the capabilities of national states and international bodies to steer national economies. However, only a fraction of financial product markets are organised and operate on a truly global basis, and even then in and through a very small number of nodes in the global economy (notably the 'global cities' of London, New York and Tokyo: Sassen, 1991). Moreover, circuits of industrial as well as financial capital have been internationalised but as relatively unco-ordinated circuits, further complicating the task of national states in managing and reproducing national economies. In these circumstances, money as a means of payment at best imperfectly represents value created through material processes. This emphasises the inherently speculative character of transactions within such an economy since credit represents a claim on surplus-value yet to be produced and there is no necessary relationship between the national territories in which it is produced and those in which claims to it are made.

Pressures 'from below' are generated because of regionalist and nationalist movements, informed by complex mixtures of cultural, economic and political motives that combine to form pressures for more powerful sub-national spaces of governance and regulation within the boundaries of national states (Anderson, 1995).[5] Economically advanced regions seek increased autonomy to reduce fiscal transfers to less successful regions (as in Catalunya in Spain and the movement to establish 'Padania' in northern Italy). Economically disadvantaged regions seek greater autonomy precisely because central state regional policies have failed to improve their economic well-being (as in Corsica and Scotland). Such separatist pressures become most powerful when economic motives combine with a sense of political oppression of culturally 'suppressed nations' (as in Quebec or in the Basque country in northern Spain and south west France). As with the shift upwards of regulatory powers, national states are not innocent and passive by-standers in these processes of territorial decentralisation of power and/or responsibilities. For example, states may seek to preserve the integrity of their national territory via granting increased autonomy to cities and regions or seek to contain fiscal crises by devolving responsibility (but not commensurate resources) for economic development to cities and regions.

These varied pressures have reinforced tendencies to shift regulatory practices from the national level and so bring about qualitative changes in relationships between national, supra- and sub-national levels. It is, however, important not to overstate the extent of such changes. There is a long history of international regulation of global economic relationships by world organisations, although the number of such organisations has expanded significantly in the last 50 years (McGrew, 1995, 29–36). The capitalist economy has *never* been one of sovereign political territories with impermeable national boundaries. There have also been growing pressures to transfer state power upwards to the supra-national level (for example, to the European Union). There has also been a long-established sub-national territorial structure to state power in response to requirements for administrative efficiency and political legitimacy. Increasingly, however, there have also been pressures further to shift the power to shape policies for cities and regions to the urban and regional levels (a decentralisation of power to decide and resources to implement decisions rather than local and regional levels simply administering central government policies for these areas) and so produce a greater correspondence between administrative spaces and the meaningful spaces of the lifeworld. As a result, more complex architectures of political power and spaces of governance and regulation have emerged.

As well as scalar shifts in regulatory capacity, there has been a change in emphasis from government to governance. Regulatory capacities have been shifted 'outwards' to non-state organisations with enhanced significance placed upon social practices beyond the state. A range of organisations and institutions within civil society has been incorporated into processes of governance, often with direct effects on the structuring of the economy. Furthermore, such formal institutions have also increasingly become organised on a transnational basis, especially in relation to environmental issues that are widely acknowledged as global (for example, Friends of the Earth and Greenpeace have both evolved to become transnational NGOs). At the same time, private sector transnational economic organisations have emerged as legitimate regulatory mechanisms: 'international standards agencies, arbitration tribunals, sectoral and functional associations and of course markets' (Radice, 2000, 734).

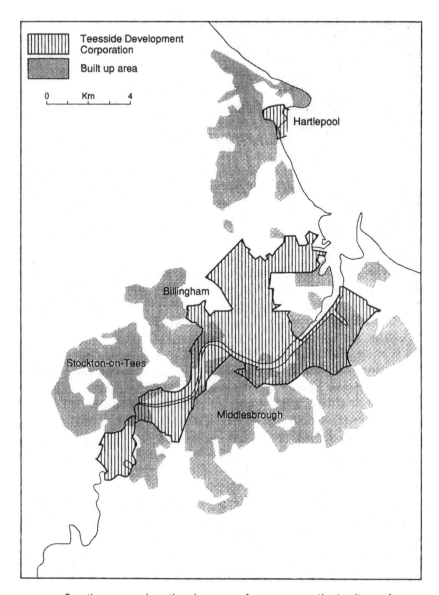

FIGURE 6.2 Creating new sub-national spaces of governance: the territory of
the Teesside Urban Development Corporation, England, 1987

In summary, the reorganisation of the national state involves moving
regulatory capacities upwards and downwards within state structures and outwards
from the state into the institutions of civil societies and back into the economic
institutions of markets. The concept of reorganisation denotes the emergence
of new, more complicated structures of regulation, involving redefined relations
between economy, society and state and complex links between scales.

Relationships between accumulation and regulation are differently constituted between and within scales. The variety of scalar territorial capitalisms finds a parallel in the variety of state capacities, strategies and spaces of governance and regulation (Weiss, 1997, 16). The growing emphasis on governance is recognition of the increasing importance – or perhaps more accurately is increasing recognition of the importance – of the institutions of civil society in securing the conditions under which the economy is possible. It acknowledges the social constitution of the economy, the embedding of the economy in cultural and political traditions and arrangements. However, this does not resolve the problems stemming from crisis tendencies in state activity but transposes them to different spatial scales and into civil society.

Some claim that the national state no longer matters. It is merely a 'nostalgic fiction' (Ohmae, 1995, 11). Others disagree, insisting that the national remains a pivotal space and scale of governance and regulation and continues to influence economic geographies. Economic globalisation is partly a product of national governmental decisions to change geographies of regulatory regimes, to create and participate in intergovernmental treaties and organisations through which national states share and co-ordinate regulatory practices, agree on common objectives and implement common regulations and standards. Moreover, the emergence of supra-national regulatory and governance institutions may further reinforce the importance of the national state (Cerny, 1990). While there has been a diminution in national state capacity to control monetary and fiscal policy in many states, these have not been uncontested tendencies (Jessop, 2000, 8–9). More generally there were and are counter-tendencies to the weakening of the national as a space of governance and the erosion of the powers of national states. Indeed, national states retain considerable power and authority in other policy domains and remain an important locus of accumulation, continuing to structure economic space. The national remains pivotal within emergent multi-scalar structures of regulation.

Even in the European Union, where the process of unbundling territoriality has gone furthest (Ruggie, 1993), national states remain significant in the governance, management and regulation of economy and society (although admittedly less so than in the past), in innovation and technology transfer (Lundvall, 1992), in environmental policy (Hudson and Weaver, 1997), and in education, training and the labour market (Peck, 1994). They also remain the supreme source of legitimacy and delegator of authority 'up' and 'down' (Hirst and Thompson, 1995). Therefore national states retain a key role as 'scale managers', shaping decisions about scalar shifts in regulatory capacity, serving as centres of persuasion and authors of narratives about change and reform and centres of interpretation and dissemination of knowledge about experiences elsewhere (Peck, 2004). The critical issue is the character of the national state, the type of regulatory regime that it maintains, and the form of capitalist economy that it seeks to encourage. Scalar changes in the forms and balance of regulatory relationships have led to corresponding alterations in national modes of regulation. There are, therefore, strong grounds for believing that national states will have a central, continuing, though to a degree different, role in processes of policy innovation, formation and implementation.

One consequence of the changing role of the national state is growing emphasis upon exploring forms of regulation beyond the market-facilitating of the liberal and the market-replacing of the interventionist, which in varying ways draw upon network conceptions of state activity and capacities. The national state has a pivotal facilitating role by encouraging and steering such policy networks. For Weiss (1997), the catalytic state is reconstituting its power at the centre of alliances formed both within and outside the boundaries of the state. Catalytic states thus seek to achieve their goals by assuming a dominant role in coalitions with transnational institutions and private sector groups. The most important partnerships are those between government and business. These processes of coalition formation are 'gambits for building rather than shedding state capacity' (Weiss, 1997, 24–7). Others emphasise the enabling state, creating partnerships between state agencies and social partners, with a broader policy agenda encompassing social inclusion (Amin, 1998). Rather than directly (or indirectly) providing goods and services on a decommodified basis *for* people, the state is now more concerned to enable social partners to provide goods and services *for themselves* with state assistance and support. Such forms of self-help are partly a response to the void left by the retreat of the interventionist state, partly a response to new and previously unmet (and unarticulated) social needs. This shift in the character of state policy is also informed by a view that more sophisticated non-price forms of economic competitiveness necessarily depend upon co-operative social relationships within cohesive and inclusive societies. However, the transition to a catalytic or an enabling mode of state activity does not mean that national states cease to have an interventionist role, any more than the transition from liberal to interventionist state ended national state involvement in the construction and regulation of markets.

6.5 Beyond and within formal spaces regulation

Within the 'formal' spaces of governance and regulation, there are interstices in which economic processes are differently and 'informally' governed and regulated. Beyond these are spaces of illegality. Four brief examples illustrate these points. First, there has been a proliferation of spaces of non-regulation. In an increasingly mobile world, there is greatly expanded movement of people in and across the non-regulated spaces of the global commons, the atmosphere and the oceans. These spaces constitute a regulatory black hole. They are beyond the remit of national states and their spaces but equally there is an absence of transnational institutions to regulate these spaces effectively (German Advisory Council on Global Change, 2002).

Secondly, there are the spaces of the informal economy. The informal economy denotes activities (for example, cleaning shoes or selling sponges) that are not illegal *per se* but are performed in ways that lie outside spaces of formal regulation. They are regulated by a variety of informal mechanisms (ranging from coercion to trust) that lie outside established legal frameworks. Sometimes this is because people who fall beyond the legally defined labour force (for example, children or illegal migrants) perform them. While the concept of informal economy was developed in relation to labour markets in the developing world, there is increasing recognition that such activities and their regulatory spaces are common across a range of capitalist economies. For example, the economies of major cities are constituted via a complex mix of formal and informal regulatory spaces.

Thirdly, there are the regulatory spaces of the illegal economy. This is the economy of illegal activities (for example, crime, drugs or prostitution: Sirpa, 2002) and/or illegal work. These spaces are often regulated by fear, physical force, and the threat of harm or death (perhaps typified by the Mafia and the Triads). Not only are the activities illegal but so too are the regulatory mechanisms. However, the boundary between legal and illegal is one that shifts over time/space: for example, in some parts of the world prostitution is legalised, while drugs that are illegal in some national territories can be legally traded and consumed in others.

Finally, there are those spaces of governance and regulation that are alternatives to the mainstream, often linked to differing definitions of value to those that prevail in the formal mainstream economy. Such activities are differently conceived from those of the 'formal' economy and are governed and regulated via different principles. Such counter-spaces and spaces of resistance, perhaps most sharply exemplified by alternative local monies such as time dollars, credit unions and local exchange and trading schemes (LETS), are more significant symbolically and politically than economically. Although limited in size and scale, they present potential radical alternatives to mainstream monetary networks and spaces (Leyshon, 2000, 442–3). As such, there may be alternative monetary networks and spaces simultaneously co-existing alongside one another and the mainstream, each with their own distinctive systems of governance and regulation. By 1999 there were over 1,300 LETS operating worldwide, for example (Williams and Windebank, 1998).

6.6 'Governmentality' and the practices and work of government and governance

The various approaches to theorisation of the state outlined above emphasise structure, institutions and spaces and it is certainly important to be able to specify the structural limits to state activity, and the territorially specific (typically national) frameworks of laws and regulations put in place to allow the economy to be possible. This territorial variability in legal and regulatory frameworks is important in influencing geographies of economies. However, such theories typically have little to say explicitly about the practices of government, governance and regulation and the ways in which these laws and regulations are implemented and put into practice. The administrative manner, style and logic by which the state regulates society in general and the economic landscape in particular as it undertakes the practical tasks of 'real regulation' (Clark, 1992) is clearly important. There is scope for a significant degree of autonomy in carrying out these tasks. For example, government officials may seek to interpret regulations 'on the borders of legality' in order to secure investment in particular spaces because of fierce inter-territorial competition or because particular locations are regarded as deeply disadvantaged or are seen as of vital political or electoral significance. In this way, the socio-political practices of implementation, the highly contested nature of regulatory spaces and the powerful actors who dominate them at a given point of time can be decisive in shaping geographies of economies.

In developing his influential and sophisticated strategic–relational theory of the state, Jessop draws together neo-Gramscian ideas as to how hegemonic practices are channelled through complex ensembles of institutions dispersed throughout civil society with Foucault's 'capillary' notion of power in theorising the mechanisms of state power and knowledge and in seeking to account for *how* state power is developed and deployed. This conceptualises power as fluid and relational, exercised from innumerable points within civil society, the economy and the state, so that many agencies and institutions are involved within productive networks of power. The Foucauldian notion of power/knowledge and concept of power as 'fluid and relational' emphasises state/civil society relations in governance. The focus is not upon who has power or the right to know/not know (that is, on a search for a single universal locus of power) but upon 'matrices of transformation' and the complex diffusion and interrelation of power throughout society. Nevertheless, Foucault privileged the role of the state (the 'macro-physics of power') as 'the point of strategic codification of the multitude of power relations (the "micro-physics of power") and the apparatus in which hegemony, meta-power, class domination and "sur pouvoir" are organised' (Jessop, 1990, 239). As such, the state is inscribed in all social relations and is simultaneously a site of power, generator of strategies and tactics and a product of strategies. Consequently, the structure and *modus operandi* of the state are historically and geographically specific.

The Foucauldian concept of 'governmentality', combining notions of 'government' and 'mentality', is helpful in further understanding these issues. For example, the emphasis on the national as the dominant space and spatial scale of regulation can be seen as expressive of one governmentality. The shift to multi-level concepts of governance, and of redefined boundaries between economy, civil society and state in processes of governance, can be seen as both indicative and constitutive of another governmentality. Not least, the spatial object of policy and spaces of governance are seen to encompass more than just the national. Government can be defined as 'any more or less calculated and rational activity, undertaken by a multiplicity of authorities and agencies, employing a variety of techniques and forms of knowledge, that seeks to shape conduct by working through our desires, aspirations, interests and beliefs, for definite and shifting ends and with a diverse set of relatively unpredictable consequences, effects and outcomes'. A mentality might be described as 'a condition of forms of thought and is thus not amenable to be comprehended from within its own perspective. The idea of mentalities, then, resembles the concept of hegemony in its emphasis on the way in which the thought involved in the practices of government is collective and relatively taken-for-granted, that is, is not usually open to questioning by its practitioners' (Dean, 1999, 11–16). As such, this raises questions about the regimes of practice through which we govern ourselves and are governed. Such an 'analytics of government' reveals 'our taken-for-granted ways of doing things' and that how we think about and question them are neither self-evident nor necessary. An analytics of a particular regime of practice, at a minimum, identifies its emergence, examines the multiple sources of the elements that constitute it, and follows the diverse processes and relations by which they are assembled into relatively stable forms of organisational and institutional practice. Such a regime gives rise to and depends on particular forms of knowledge and, as a consequence, becomes the target of various programmes of reform and change. The disciplinary knowledges of the social sciences play an important role in providing an 'intellectual machinery' of ordering procedures and explanations that construct and frame reality in ways that allow governments to act on it (Rose and Miller, 1992).

Governmentality emphasises the practices, the 'how' questions, rather than the structures of government and governance (although it is important not to forget or to underplay the significance of these structures). Governmentality 'is intrinsically linked to the activities of expertise, whose role is not one of weaving an all-pervasive web of 'social control', but of enacting assorted attempts at the calculated administration of diverse aspects of conduct through the countless, often competing, local tactics of education, persuasion, inducement, management, incitement, motivation and encouragement' (MacKinnon, 2000, 296).[6] Moreover, such activities are territorially demarcated. Space is an important element of governmentality because 'to govern it is necessary to render visible the space over which government is to be exercised. And this is not simply a matter of looking: space has to be re-presented, marked out' (Thrift, 2002, 205). This thereby locates the space of the state as one element in wider circuits of power and moves from a position that sees the state as simply an explanation of other events to one that regards the specific activities of the state as themselves to be explained and the black box of the state as requiring opening up in order to explain how it can perform with a degree of functional coherence. Such internal coherence can only be achieved through the successful realisation of specific 'state projects' which unite state agencies and officials behind a distinct line of action (Jessop, 1990, 229). Achieving such unity is a contingent matter. Even if it is achieved, however, there is no guarantee that such projects will always and only have their intended effects. An emphasis on competing tactics and unintended consequences resonates with more general arguments about emergent effects and with ideas of crisis theories and a structural tendency to a crisis in crisis management precisely because of the inability to anticipate all the impacts of action.

This has several significant consequences. The first relates to the constitution of the objects, subjects and spaces of government. For example, national and regional economies are constituted via territorial statistics, industries by industrial statistics. Such statistics have a key role in making economies visible and constituting them as objects for policy action. Secondly, Latour (1987, 237–40) emphasises the key role of 'centres of calculation', critical nodes in which information on distant objects is brought together, compared, combined, and aggregated via use of mathematical and statistical techniques, thereby enabling government to 'act at a distance' on objects of its programmes and policies. Thirdly, it highlights 'the specific mechanisms, procedures and tactics assembled and deployed as particular programmes are materialised' (MacKinnon, 2000, 295) and through which governmental programmes are activated and put into practice. Particular techniques and practices become governmental because they can be made practical, transformed into concrete devices for managing and directing reality. Inscription (for example, writing down agreed quantitative targets) and calculation are key technologies, enabling 'enclosure' to be breached by 'responsibilising' and disciplining actors to the claims of central authority. These technologies render reality 'stable, mobile, comparable, combinable', enabling government to act on it (Rose, 1996). For example, the turn to neo-liberalism generated new technologies, deployed to implant new modes of calculation by creating simulacra of markets within the public sector. This move to government via pseudo-markets was closely linked to the transition from government of society (via the welfare state) to 'government through community' and 'community' as a new governmentality. Another example relates to the tensions between decentralisation to sub-national spaces and the development of new managerial technologies at national level to 'steer' the activities of local and regional agencies and ensure that they deliver national policy objectives. Sub-national spaces become simultaneously objects and subjects of national government, and via 'the combination of flexibility and standardisation (that is, different levels, same targets) ... gives governmental technologies their utility as instruments for managing space' (MacKinnon, 2000, 309).

6.7 Summary and conclusions

Without effective processes of regulation, governance and governmentality, economies could not be constituted and reproduced. Capitalist economies are grounded in competitive relations that must be kept within agreed and accepted bounds to allow the economy to be performed and the accumulation of capital to proceed more or less smoothly. Historically, such processes of regulation were strongly associated with national states and with the bounded space of the national territory. The extent and scope of national state activities has altered, especially over the last quarter of the twentieth century, as national states have been 're-organised', shifting regulatory capacity both upwards and downwards to other spatial scales and outwards from the state back into markets and civil society. These changes partly reflected the increasing internationalisation of critical circuits of capital, partly increasing fiscal pressures on national state budgets, and partly political ambitions and forces seeking to operate at scales other than the national. Nevertheless, the national remains an important regulatory scale, the national state an important regulatory institution.

While the economy remains predominantly constituted in and through scaled territorially-bounded spaces, the expansion in networked forms of economic and social organisation with different spatialities cutting across these boundaries is posing challenges to such territorialised spaces of regulation. This is creating tensions between territorially defined modes of governance and regulation – and the rights and responsibilities of citizens of these spaces – and the prosecution of economic, political and social interests within 'aterritorial' forms of networked relations. However, there are considerable powers of resistance vested in such spaces, powerful interests that have no wish to see their capacity to govern and regulate diminished any further. Moreover, relationships between territory and networks are complex, as to some extent national states have encouraged and sanctioned networked forms of international regulation and, in some instances at least, the powers of national states (especially the most powerful, the USA) have been underpinned by these developments. The net result is that systems of regulation have redrawn the boundaries and links between economy, state and civil society, become more multi-scalar, and combine different spatialities of governance.

Notes

1 However, the state is not the only national scale regulatory institution and mechanism and the national is not the only territorial space and scale of regulation (see below).
2 This variety makes it more appropriate to refer to a Regulation School (Dunford, 1990) rather than a regulation theory.
3 'Hollowing out' is an Euro-centric concept, as state power has been shifted to supranational EU and regional levels. Elsewhere, constructing the national as an effective regulatory space is seen as imperative while other states, notably the USA, maintain enormous structural power.

4   Chesnais (1993, 12–13) suggested that only Germany, Japan and the USA were not being undermined in this way. As of 2003, this probably only holds true of the USA.
5   Such tendencies were not universal. In the United Kingdom in the 1980s, for example, there were strong centralising tendencies at national level.
6   However, concentrations of expert knowledge can unintentionally give rise to 'enclosures', tightly bound sites of vigorously defended professional expertise, resistant to the wishes of government (Rose, 1996).

7 Spaces of Production

7.1 Introduction

The 'moment of production' is critical within the circuit of capital and the reproduction of the social relations of capital. Production cannot occur everywhere but must occur somewhere. In this chapter the focus is on those defined spaces in which commodity production occurs within the formally regulated economy, particularly production of material commodities and processes of appropriating and nurturing nature that have increasingly been transformed to mimic those of transformative production in manufacturing. The spaces in which employed labour works are central to the production process, as is the recruitment of appropriate labour and the regulation of the labour process to ensure the production of surplus-value. Equally, the ways in which various spaces of work are combined into social systems of production are of crucial importance. In the latter part of the chapter I briefly consider non-mainstream forms of production within capitalist economies.

7.2 Spaces of work and the organisation and regulation of the labour process

The collective and social character of production necessitates assembling workers in specific spaces of work at defined times, to allow discipline to be maintained. Such a space constitutes a system of labour control, surveillance and resistance. Managers and workers negotiate and struggle around the 'frontier of control' that runs through it.[1] Workers strive to maximise their remuneration and control over the labour process and optimise working conditions. Managers endeavour to guarantee the production of surplus-value[2] by organising the labour process, the working day and social relations so that workers are imbued with desired attitudes, beliefs and values. In this way they seek to ensure that authority relations appear at best just, or at least inevitable. However, workers may contest these 'enabling myths' (Dugger, 2000). Managers may then have to adopt other, non-discursive tactics to secure the frontier of control.

7.2.1 Craftwork and flexible specialisation

Craftwork requires that workers have considerable control over the process and pace of work, with extended work time cycles. The knowledge and skills of craft workers cannot be disembodied and transferred to machines. Consequently, the labour process must be controlled by (neo)paternalistic strategies of co-operation and 'responsible autonomy' (Friedman, 1977), encouraging workers to identify and ally their interests with those of their company and connect spaces of work with life in the community and home.

While there has been a long-term tendency for capital to replace skilled by less skilled labour, craftwork nevertheless remains of critical importance (Pollert, 1988), especially in some types of firm (notably SMEs), activities and sectors. The knowledge and skills of craft workers have been emphasised as critical to the success of 'flexible specialisation' in mature industries in industrial districts and also in newer 'high tech' industries as technological change creates new skills, specific to particular workers (Storper, 1993). Workers in SMEs performing knowledge-intensive activities in 'high-tech' industries such as electronics work at self-determined speeds within relatively non-hierarchical managerial structures. Furthermore, 'creative' occupations in SMEs producing and transmitting various forms of knowledge have expanded greatly. Perhaps proto-typically, 75% of workers in software companies and in new semi-conductor firms in the USA comprise professional-managerial and technical staff (although this also reflects out-sourcing of fabrication). Consequently, 'reflexive' jobs characterised by long job-cycles, ranging from days to weeks (engineers and technicians) to months or years (advertising executives) have grown in number. Such 'soft capitalism' (Heelas, 2002) can, however, lead to super-exploitation, working very long hours (far beyond those to which people are formally contracted), driven by commitment to the company and personal satisfaction from performing their job (Massey et al., 1992).

SMEs reliant upon craftwork or deploying 'flexible specialisation' approaches pursue selective, often spatially specific, recruitment and retention policies. Recruitment is often via personal ties, family and friends, depending on 'who you know' as well as 'what you know' or what you can do. Companies producing commodities with a high symbolic content, such as advertisements, recruit on the basis of reputation, emphasising reliability for more junior positions, creativity for more senior ones (Grabher, 2001, 369). Competitive strategies based upon flexibility of production or product quality require long-term retention of skilled workers, socialising them into the company's culture, as their acquired practical knowledge and capabilities become key corporate competitive assets. Company-specific skills and knowledge cannot be purchased in the labour market (Penrose, 1957).

7.2.2 Mass production and Taylorism

The American System of Manufacture enabled parts to be machine-processed and assembled 'by workers who had not been apprenticed in the craft tradition'

(Best, 1990, 32). However, the full implications of this development for organising the labour process remained latent until Taylor's work on scientific management and Ford's mass-production automobile plant with its mechanised assembly lines. Mass production requires a profound separation of mental from manual work, extreme task specialisation and a deep technical division of labour. Alienated workers perform simple, repetitive deskilled tasks with very short job-task cycles (often defined in seconds). The moving production line delivers materials to them at speeds determined by management. Increasing line speeds as a result of managerial dictat leads to labour productivity increases while shift systems allow maximum utilisation of machines.

Companies using mass-production strategies typically seek out spaces with abundant unskilled or semi-skilled labour. They recruit and retain selectively if labour market conditions permit. For example, Ford recruited very selectively at Highland Park in the USA in the early twentieth century and at Halewood on Merseyside some 50 years later (Beynon, 1973). However, Taylorist labour control strategies allow companies to switch to less selective approaches. Since workers require little training for routine production, companies can pursue aggressive 'hire-and-fire' tactics, often with rapid labour turnover. While mass production initially conferred great competitive advantage, it subsequently became more problematic. The initial challenges arose from increasing resistance by workers to its Taylorist command-and-control culture in 'full employment' conditions in the advanced capitalist countries. Increasing resistance to speed-up and intensification of work was expressed in numerous disruptive official and unofficial 'wild-cat' strikes in the 1960s. Later, more fundamental challenges were posed by increasingly volatile product markets.

Despite claims about the demise of mass production, Taylorism is far from dead. First, Taylorism has been extended into agriculture and mining, activities characterised by labour processes of 'appropriation' and 'nurturing nature', which often necessarily cede considerable autonomy to workers. This was perhaps *a fortiori* in underground coal mining. The separation of workers from managers led the latter to seek to restructure the labour process to resemble that of the factory production line (Winterton, 1985) and shift production from deep to strip mining and labour processes more amenable to surveillance and control (Beynon et al., 2000, 16–35 and 64–89). In agriculture, there are attempts to 'outflank nature' by industrialising and Taylorising production. Secondly, Taylorist principles have been introduced into many routine sales and service activities, creating 'downgraded services' as work is reorganised to cut costs and tighten managerial control of the labour process. Thirdly, the principles of Taylorisation have been further extended with the expansion of 'downgraded manufacturing' (Sassen, 1991). Fourthly, mass production remains prominent, not least in technologically dynamic 'high tech' computing and electronics sectors in which scale economies are crucial (Delapierre and Zimmerman, 1993, 77–8).[3] Fifthly, the growing size of corporations enabled them to split up production processes, functionally and spatially, while technological changes

FIGURE 7.1 Automating production in the steel industry, Teesside, England

allowed deskilling of jobs. Consequently, companies could shift deskilled work to spaces with large masses of people prepared to undertake such work, usually for lower wages. Geographies of Taylorist mass production were thus recast at various scales, stretching the social relations of production over successively greater distances (Lipietz, 1987). Such 'spatial fixes' (Harvey, 1982) enable Taylorist manufacturing to be preserved by (temporarily) containing its inherent contradictions within new intra-national and international divisions of labour. Indeed, relocation can enable companies to continue to use despotic employment practices akin to 'bloody Taylorisation' (Lipietz, 1986). Finally, another, sometimes linked, neo-Fordist (Palloix, 1976) response was to increase automation (for example, by using robots in production: Aglietta, 1979, 123) to alleviate problems of control over the labour process. As such, neo-Fordism can be seen as a further attempt to perfect flow-line principles within mass production.

7.2.3 High-volume flexible production strategies

High-volume flexible production (HVFP) strategies seek to combine the positive aspects of both mass and craft production while avoiding the problems of both. In general HVFP involves (relatively) small batch production from mass-produced components. While developed in the automobile industry, it has subsequently diffused, often in hybrid form, to other consumer goods industries (for example, personal computers) and to bulk continuous flow industries (such as chemicals and steel). At its limit, HVFP produces mass-customised commodities, unique products manufactured from mass-produced components and parts (Pine, 1993). While often represented as a radical departure in terms of production organisation, however, many complex, specialised and sophisticated products, such as major items of capital equipment (power stations, generators) and complex commodities (large passenger ships), have always been produced on this basis (Vaughan, 1996).

Because HVFP typically incorporates principles of 'just-in-time' and 'lean' production, predicated on minimal stock levels, companies carefully recruit individual workers and organisations to represent them. They require specific forms of non-adversarial, compliant, company-oriented trade unionism or company consultative councils rather than unions. Workers are carefully selected, often using extensive psychological and physical dexterity tests, to ensure trouble-free and error-free production, with the desired characteristics of workers varying between industries. For example, in the automobile industry, companies seek young, physically fit, mainly male workers who eschew disruptive industrial action. In electronics, in contrast, companies in countries such as China, Mexico and Thailand seek to recruit young women, often single mothers prepared to accept poor working conditions in order to provide for their children (CAFOD, 2004). The precise form of recruitment, retention and labour control strategies depends upon labour market conditions. For example, in tight labour markets (such as Japan from the 1950s to 1990s) promises of 'jobs for life' for core workers, perhaps in exchange for the possibility of redeployment within the firm, were pivotal. In labour markets characterised by high unemployment, masses of people seeking work and many applicants for each available job, companies (often the same companies) can be more selective in their recruitment criteria (seeking out 'green labour') and in their retention strategies (CAFOD, 2004; Fucini and Fucini, 1990; Hudson, 1995).

Echoing responsible autonomy strategies, HVFP requires management of the labour process through employee commitment to and involvement in the job via Human Resource Management (HRM) and Total Quality Management (TQM) practices (Wills, 1998, 15–16). There is emphasis on quality enhancement, problem solving rather than machine minding, and longer job-task cycle times, built around (sometimes conflicting) themes such as flexibility, responsibility, self-development, job enrichment, training, security of employment, performance-related pay schemes, teamwork and improving the commitment and trust of a more valued workforce. However, companies practising HVFP strategies require great flexibility in allocating workers' time on the line, considerable use of multi-tasking, increased flexibility in the scheduling of overtime, and reorganisation of shifts to ensure that factories and machines produce goods for the maximum time possible within regulatory limits.[4] Consequently, the intensive pace of work leads to physical injuries, such as repetitive strains (Leslie and Butz, 1998).

Relative autonomy at the point of production is contingent upon accepting stringent productivity and quality targets and systems of managerial control.

Requiring production-line workers to perform their own maintenance work erodes the distinction between craft and assembly line work. Enhanced managerial control over the labour process requires detailed information about individual workers' performance. As such, it is 'important to counterpose the contemporary business rhetoric and practice of targets, ranking and assessment' with the 'new management ideals' of HRM (McDowell, 2001, 237). Furthermore, defining internal workplace social relations around customer–supplier relations (Yates, 1998, 127–37) creates a system of labour control that individualises work norms and remuneration systems within a culture of teamwork and competitiveness. Replacing direct supervision with peer pressure helps redefine workers' identities and fracture their sense of solidarity (Hudson, 1997). Companies intensify the pace of work by new, subtle methods of control, exploitation and surveillance, increasing psychological stress on workers (Okamura and Kawahito, 1990), while cutting workforces and the turnover time of fixed capital. As such, HVFP represents a logical and historical extension of the principles of mass production and has become increasingly prevalent (Economic Policy Institute, 2000).

7.2.4 *New ways of producing, new forms of labour regulation and 'soft capitalism'*

The sense of satisfaction from doing a job well has been emphasised in more celebratory accounts of working within systems of HVFP, contrasted with the alienating effects of Taylorist mass production, re-engaging workers creatively with the production process.[5] The sense of satisfaction from a job well done is also an important attribute of craftwork, helping confer and create personal identity. This emphasis upon personal satisfaction from and development through new forms of work organisation has been elaborated within discourses of 'soft capitalism'. Work continues to have a strong utilitarian aspect (unavoidably, as required goods and services must be purchased) but this is only part of the story. Soft capitalism is 'about culture, knowledge and creativity; about identity, about values, beliefs and assumptions' (Heelas, 2002, 82), exploring 'cultural expertise concerning the psychological realm of life' and developing it for commercial and competitive advantage. This leads to a specific work ethic. Work provides 'the opportunity to "work" on oneself; to grow; to learn; to become more effective *as* a person'. By working on oneself through work (in particular via training and supervision) 'one becomes a more efficient producer. But motivation to work is also – crucially – enhanced by virtue of the fact that *personal* development is involved. Work is meaningful because (among other things) it provides opportunities for "inner" or psychological identity exploration and cultivation'. Thus 'work *really* matters, or matters *most*, when it caters for what it is to be alive. Without the training, without the opportunity to *be*, *explore* and *develop* oneself, work is deemed to be unsatisfactory' (Heelas, 2002, 93).

Indeed, at the limits of soft capitalism, work acquires a spiritual dimension: 'one is working on oneself to experience that spirituality which is integral to one's very nature or essence. The workplace is valued ... as a vehicle to the end of self-sacralisation. And inner spirituality can then be put to work to enhance work productivity' (Heelas, 2002, 89). Clearly, issues of productivity and surplus-value production are never far from the surface in such discourses, despite the references to self-improvement, spirituality and the sacred. Even so, the extent to which such a work ethic, centred on 'life values' and a 'life ethic', is generalisable within the constraints of capitalist relations of production is debatable. For work 'is increasingly insecure and stressful, for managers as much as any other group and ... the encouragement of personal commitment is little more than rhetoric for a small proportion of managers' (Warde, 2002, 189). While work for a small cadre of managers may become 'meaningful', it is difficult to see how much routine work could be so transformed. Indeed, the expansion of Taylorisation into new spheres suggests that much work has become less, rather than more, meaningful.

There are, however, also questions as to the regulation of work within the parameters of 'soft capitalism', as this lays great weight upon personal self-discipline as well as personal development. The practices of self-discipline may enhance (self) exploitation and surplus-value production rather than spirituality, benefiting capital rather than labour since the most positive and effective disciplining of individuals is achieved through the practices of freedom (Rose, 1999). For capital to be able to hand over the management of labour to people for whom such self-management is increasingly understood as constituting 'pleasure in work', and the development of the self, is to achieve unprecedented and sophisticated levels of regulation of labour and the labour process (Donzelot, 1991). As such, the practices of soft capitalism take processes of self-regulation of work to new heights, suggesting that for labour, at least, these would be better characterised as 'hard' not 'soft' capitalism.

7.3 Linking spaces of work and firms into social systems of production

Broadly speaking, there are three ways of linking spaces of work into social systems of production. Companies can produce in-house, form arms-length market relationships with other firms or construct network relationships based on longer-term collaboration and commitments to other companies. A firm must therefore decide to make, buy or collaborate to ensure production. However, no company of any significance has been wholly vertically integrated while forms of external linkages have varied over time/space. With the growth in significance of collaboration, firms become mechanisms to co-ordinate inter-firm systems of production of varying degrees of socio-spatial complexity, encompassing many legal entities (Dicken and Thrift, 1992, 285). Specifying the boundaries of the firm has become increasingly difficult as companies confront fluctuating choices at the shifting boundary between vertical integration and disintegration (Veltz, 1991, 210–12). This dilemma is resolved in varying and contingent ways, depending on the balance of advantages between stabilisation of inter-firm relations and the 'significant degrees of freedom' within networks and the asymmetries in power relations between firms.

7.3.1 *Firms as systems of production: managerial models and spaces*

Vertically integrated firms emerged in the USA around the end of the nineteenth century, with production relying upon intra-company capabilities and product competencies, co-ordinated by a firm's internal hierarchical management organisation. Companies produce in-house for various reasons. It may be cheaper and/or more certain, minimising risks. Particular tasks may be too exacting and specialised to be sub-contracted or involve firm-specific knowledge that must be protected from competitors. Vertically integrated production may entail producing components at differing locations within an intra-corporate geography of production, depending on the relative weights attached to economies of scale and scope. Irrespective of their dispersed geographies, however, decisions about production organisation remain within the internal corporate control hierarchy.

Despite recent emphasis upon growth in external sourcing, there are persistent tendencies towards vertical integration, not least in 'leading edge' sectors such as computer software (Coe, 1997). In semi-conductor production, fixed capital costs have increased as a proportion of total production costs. Such sunk costs make 'entry to and exit from the industry more expensive and difficult' (Flamm, 1993, 66–70). Furthermore, the tendency to blur boundaries between material products and services is encouraging vertical integration, with companies extending activities along the supply chain (for example, in automobiles: Hudson and Schamp, 1995). In industries ranging from computers to steel production wholesalers are extending their influence both up-stream and down-stream. They are responding to the impacts of just-in-time production and market fragmentation, leading to demands for greater flexibility in production, by customising products to customers' demands (Hudson, 1994c).

Companies reshape their internal spaces of production and management via process, product and organisational innovation.[6] The American System of Manufacture required creating impersonal bureaucratic management control methods, encompassing new middle management tasks and managerial structures. With the emergence of industrial corporations in the USA by the early twentieth century, the dominant managerial form became a single hierarchy within a single corporation. This replaced market co-ordination as the means of managing geographically scattered and diverse business and production activities (Best, 1990, 35–46). These vertically integrated systems were, however, unwieldy. The paradigmatic response to this problem was the multi-divisional managerial structures developed at General Motors, based around two central guiding principles (Ghosal and Bartlett, 1997). First, decentralisation to increase the number of employees who could exercise entrepreneurial judgement. Secondly, new cost accounting systems to co-ordinate entrepreneurial initiatives and maintain overall corporate coherence. Other major companies (including Du Pont, IBM and Philips) subsequently adapted this model.

From the early 1970s, the 'command and control' model – whether in hierarchical or decentralised multi-divisional form – became increasingly problematic (Pasternack and Viscio, 1998). Companies sought new, more decentralised management spaces and structures appropriate to 'flexible production', increasingly volatile market conditions and enhanced emphasis upon product differentiation and market segmentation. Consequently, enabled by advances in production and communication technologies, the 'newly emerging organisational form' is 'the complex global firm' which has as its 'key diagnostic feature' an integrated network configuration and capacity to develop flexible co-ordinating processes both inside and outside the firm (Dicken et al., 1994, 30). This creates scope for greater variety in internal organisational structures, governance mechanisms and geographies: internal organisational forms become increasingly heterarchical (Grabher, 2002).

Based on the canonical example of ABB (Asea Brown Boveri), corporate success became seen to depend upon people, their collective and combined knowledge within a company, and the coherence with which different parts of the company are combined within looser, more decentralised and fluid horizontal network structures of management. Companies were typically split into separate profit centres, with careful central monitoring of their performance. For example, ABB's Abacus Management Information System collects data for 4,500 profit centres and compares performance and budget forecasts on a monthly basis (Taylor, 1991). The initial model was subsequently modified by ABB twice in the late 1990s, following the scrapping of the much admired and imitated matrix management structure in 1998. The latest changes, focused on customer and product segments, are 'a natural progression of the journey we have been on', a response to a 'silent revolution' in the market that is 'completely changing the business landscape'. ABB seeks to transform itself into an 'agile knowledge-based company, with "brain power"' as its corporate motto, the Internet as its favourite tool, allowing it to interact one-to-one with customers and deliver mass-customised information, products and services. However, such highly flexible mass customisation requires common business and management processes worldwide to facilitate ABB's interaction with its customers (Chief Executive Jorgan Centeram, cited in Hall, 2001). It has also been modified in various ways by other companies, but with an emphasis on decentralisation and loose horizontal relationships. For example, GSK (GlaxoSmithKline) claimed to be creating an organisational structure for

R&D 'completely different to anything that exists today', disaggregating research into individual profit centres involved in intra-corporate competition for funds (Pilling, 2000).

In the contemporary globalising economy large companies seek to capitalise on their carefully selected core competencies, devising new 'hybrid' dual organisational structures and 'strategic architectures' to cope with the complex challenges posed by increasingly fluid and dynamic markets (Howells, 1993, 222). These have two components. First, an integrated network structure to manage core competencies, encompassing intense knowledge exchanges within the firm. Decisions about core competence activities require long-term commitments based on sunk costs and specific governance mechanisms to ensure the effective circulation of knowledge (Amin and Cohendet, 1997, 13). Secondly, there is a classical hierarchical managerial structure to manage regular, ongoing, non-core competence activities. This is needed for two reasons. First, it is a cheaper way of managing routine activities. Secondly, non-core activities principally require regulation of flows of codified information. Whether such activities are retained within the firm depends upon 'traditional' transaction cost criteria. There is thus a major fault line between core and non-core activities but this line – and so the boundary of the firm – is fluid and shifting. Indeed, it may be more helpful to think of a shifting continuum of activities than a core/non-core dichotomy.

In some circumstances, however, firms are developing more differentiated internal management spaces and structures and more decentralised managerial strategies as they seek to respond to local opportunities. Such heterarchic firms – often producing commodities with a high knowledge and/or symbolic content – are characterised by organisational heterogeneity. There is a necessary degree of 'under-determination' or 'under-specification' in production via 'soft assembly'. The form of the final product is not obvious at the start of the process and emerges via search and experiment. This requires a

> redundancy of resources, skills, models and philosophies ... embedded in different organisational contexts and in different organisational layers in a way in which the higher layers subsume the activity of the lower organisational layers by controlling it only in a limited way. Such 'soft assembly' allows lower organisational units to respond to local contexts and exploit intrinsic dynamics: 'soft assembly' out of multiple, largely independent components yields a characteristic mix of industries and variability. (Grabher, 2001, 358)

This tendency is related to the emergence of a 'more reflexive and 'soft' capitalism, an 'autopietic system' characterised, *inter alia*, by four features (Thrift, 1999, 67). First, an interlocking institutional arrangement of business schools, management consultancy and management gurus. Secondly, the growth of the business media. Thirdly, the growth of new business practices, including HRM, marketing and a growing interest in leadership, communication and other 'soft' skills. Finally, 'simply momentum', a self-reinforcing fashion effect. Consequently, expertise and expert knowledge, and its diffusion between and within firms, are central to the emergence of new managerial models and styles. However, 'there is nothing new about the "soft" dimension with regard to economic activity'. What *is* relatively new is 'the *degree* to which culture is called on; the *degree* to which experts exercise their judgement; the *degree* to which creativity is called into play' (Heelas, 2002, 81, emphases in original).

The net result has been growing emphasis on the need for flatter, more flexible, differentiated and decentralised managerial structures and corporate cultures, for a new mode of corporate governmentality within an increasingly risky, unstable and rapidly changing economic environment. Allegedly, 'firms now live in a permanent state of emergency, always bordering on the edge of chaos' (Thrift, 2002, 201). Consequently, they become faster, more agile, generating just enough organisational stability to change in an orderly fashion and sufficient hair-trigger responsiveness to adapt to the 'expectedly unexpected'. This in turn is creating demands for new managerial skills and 'fast subjects', a qualitatively different type of manager able provide leadership and creativity while living life in a blur of change. Consequently, there is an attempt 'to engineer new kinds of fast subject positions which can cope with the disciplines of permanent emergency.' This project 'involves much more direct engineering of the management subject' than has previously been attempted. Further, and crucially, it involves 'the production of new spaces of intensity in which the new kind of managerial subject can be both created and affirmed. This is a project, in other words, which relies on the construction of an explicitly geographical machine'. Creating such 'fast' management subjects 'requires finding a whole variety of different methods of acting upon others in order to produce subject effects' (Thrift, 2002, 202–5). This entails specifying procedures which both reveal and value the 'new things' that are needed to create 'fast subjects', which Thrift (2002, 206–7) 'rather glibly' names as *sight, cite* and *site,* thereby seeking to emphasise 'how bound up these new procedures are with the production of new spaces which, by being more active, more performative than those of old, can help foster creativity. Thus sight – new spaces of visualisation; cite – new spaces of embodiment; and site – new spaces of circulation.' The latter are of particular relevance since new means of producing creativity and innovation are bound up with new geographies of circulation, intended to produce situations and spaces conducive to creativity and innovation.

Nevertheless, 'it would clearly be ludicrous' to suggest that the project of governmentality centred on the spaces of production of fast subjects 'has a total grip' (Thrift, 2002, 207). Indeed, it remains an open question as to whether such a project is feasible. This 'fast world' may not last, running up against its limits and the contradictions of underlying capitalist social relations. Not least, this is because it is grounded in a particular conception of capitalism and a business model centred on short-termism that came to dominate in the USA in the 1980s and 1990s but now (post-Enron, Parmalat, WorldCom and other accounting and financial scandals) looks increasingly fragile.

7.3.2 Constructing spaces and systems of production via inter-firm relations

7.3.2.1 RELATIONSHIPS GOVERNED BY PRICE AND MARKET RELATIONS

Production systems can be co-ordinated via market relations: customers specify the required technical and performance criteria, suppliers then bid for orders. Market regulated supply relationships are relatively distant and non-interventionist. Pricing policies reflect and shape relations between companies depending, in part, on the character of commodities, markets and links between customer and suppliers (Leborgne and Lipietz, 1991). Price competition is often the strategy adopted for supplying components and parts embodying little specific knowledge or relatively simple commodities mass produced via labour-intensive processes for final consumption (such as shoes or clothing).

Supply chains are often structured so that smaller and weaker firms are confined to market niches that are unattractive to larger and more powerful firms and on terms that favour the interests of the latter. Often companies sub-contract opportunistically in response to short-term fluctuations in demand. Such sub-contracting can take many forms: for example, 'putting out', sub-contracting work to smaller companies and home-workers, and 'splitting in', with external sub-contractors bringing their workers into the factory to work on the contracting company's machinery (Hadjimichalis and Vaiou, 1996, 7). Local labour market conditions may be important in deciding whether to sub-contract. For example, in the absence of (militant) unions, large firms may have no incentive to sub-contract work to low-wage secondary labour market firms (Manzagol, 1991, 214–15).

7.3.2.2 NETWORKS, RELATIONAL CONTRACTING AND THEIR GEOGRAPHIES

Relational contracting occupies a 'middle ground' between the 'tight dualism' of firms and markets (Sayer and Walker, 1992, 128–9), involving ongoing relations of exchange, interaction and mutual development between two or more firms. Such contracting typically develops in situations in which one company relies on another for parts, components or services embodying considerable skill and requiring trust in sharing knowledge and key competencies (Schamp, 1991). Systemic sub-contracting is well established in industries such as shipbuilding but in others – computers, telecommunications and other 'high tech' electronic equipment – it has increasingly become central to production strategies (Hudson, 1989). As such, this growth offers potential for firms of all sizes to engage in integrated network activities. Moreover, a corollary of such deeper relationships is that the supplier population becomes increasingly differentiated. Many major firms now rely on an upper tier of suppliers, closely integrated at all stages of the production process, from design to final production

(Dicken et al., 1994, 39). These new, closer and longer-term customer–supplier relationships exhibit a greater degree of embeddedness, predicated upon non-market forms of interaction and mutual trust.

Relationships of 'trust' are thus both produced through and help further reinforce close inter-firm relations (Gertler, 1997, 47–8), especially in situations of repeated interaction in a defined space (Crewe, 1996). Building trust requires the use of mutually understandable explicit language and often prolonged social-isation or two-way face-to-face dialogue that provides reassurance about points of doubt and leads to willingness to respect the other party's sincerity (Nonaka and Takeuchi, 1995). Creating trust has three main benefits. First, being able to rely on others saves time and effort. Secondly, trust reduces risk and uncertainty, and reveals possibilities for action that would otherwise be concealed. Thirdly, it expedites learning because people and organisations are privy to richer and thicker information flows and people divulge more to those they trust. However, even if transactions *are* grounded in relationships of trust, these are not *neces-sarily* transactions among equal partners. For example, 'open book' negotia-tions, with companies exchanging information about costs, orders, investment plans and competitors, often require suppliers continuously to reduce prices. Moreover, the concept of trust is problematic (Lane and Reinhard, 1998), open to competing interpretations. 'Trust' has very different meanings in the context of mining coal underground (Douglass and Krieger, 1983) and that of produc-ing complex financial products, such as derivatives designed to hedge risk, in the City of London (Budd and Whimster, 1992). However, in both cases it results from day-to-day practices, strongly conditioned by surrounding social institu-tions and regulatory regimes. As such, 'trust' must be (re)produced, perhaps from mistrust, and protected when endangered by opportunism, to allow people to have confidence in the probity of others.

Relational contracts often involve complex configurations of several firms, linked into networks of varying form. These can be categorised in terms of degree of closure/openness and the extent of hierarchy within the network. Networks vary from relatively closed (for example, Japanese *keiretsu*) to much more open-system networks. They also vary from relatively egalitarian hori-zontal associational networks, such as those found in parts of the Third Italy (Asheim, 2000), to vertically differentiated sub-contracting networks, incor-porating varying degrees of power and influence and hierarchical and exploitative relations among firms (Leborgne and Lipietz, 1991, 38–9). Network relationships are thus constituted in diverse ways, with varying degrees of legal (in)formality and differing structures of intra-network power relationships. The location of decisive power depends upon the character of the commodities produced and the cultures of production in inter-firm rela-tions.[7] In food production, for example, major food retailers have become increasingly sensitive to consumers' concerns about food quality (Marsden et al., 2000). Consequently, they dictate the parameters of animal rearing and crop production as these concerns are transmitted down the supply chain. In like manner, McDonald's stringently specifies how its franchisees should cook and prepare hamburgers. In manufacturing, assembly companies specify precise quality standards and delivery schedules to which their component suppliers must conform (Hudson, 1994a). In mass-consumption industries the retailer–manufacturer link may crucially shape the production system and its geography. Hourly variations in demand can be directly relayed to the production line from electronic points of sale (EPOS) in shops via electronic data interchange (EDI). For example, in the clothing industry 'the need for close monitoring of suppliers who can make the quality grade and respond extremely quickly to fragmenting and shifting consumption signals is resulting in more localised sourcing chains' (Crewe and Lowe, 1996, 279). Major clothing retailers pass the costs of stockholding on to suppliers by securing 'pseudo just-in-time' supply from increased stocks held by producers (see also Hudson and Schamp, 1995, for similar arrangements in the automobile industry).

The precise configuration of links and relationships depends upon the character of the production system. For example, HVFP systems require and reflect a particular conception and practice of inter-firm co-operation, to ensure that rigorous demands for quality and delivery schedules are met and that suppliers accept a greater share of R&D and product development costs, in exchange for long-term but never unconditional contracts. Within HVFP systems, major assembly companies increasingly select a few first-tier suppliers, which function as 'system integrators', assembling components produced by others into sub-assemblies and final products – echoing a long-established model of organisation in 'one off' or small-batch production of capital goods. While a few upper-tier suppliers (often major multinationals) may enjoy relationships based upon trust, further down the supply chain regulation via price competition remains endemic. Suppliers have been increasingly subject to rigorous screening and selection procedures by their customers while renewal of contracts has typically depended upon price reductions, passing on productivity gains resulting from co-operative R&D and continuous quality improvement to final product manufacturers (Hudson, 1994a). Similarly, in networks dominated by major retailing companies, such as clothing, the production system is structured to lock preferred suppliers into relations of dependence upon oligopolistically organised retailers, with fierce price competition between 'preferred' and lower-tier suppliers (Crewe and Davenport, 1992, 196) and, on occasion, between retailers and 'preferred' suppliers (Minton, 2000). Despite their relatively flat nature and collaborative character, there are significant differential power relations within networks. Not all parts of a network are equal (Dicken et al., 1994, 32).

In other networked production systems, such as those producing advertisements, lead companies perform co-ordination in different ways, depending upon the character of the inputs that are out-sourced (Grabher, 2001, 367). In co-ordinating the provision of routine services, such as printing, for which there is little creative input and a familiar pattern of 'made-to-order' supply, advertising agencies act more as 'orchestrators'. For the provision of other commodities, which cannot be pre-specified and have a high creative input (such as film direction and production), a jazz improvisation metaphor is more appropriate.

For 'orchestration connotes prescribed musical scores and a single conductor or leader to create the static (hierarchical) synchronisation of the orchestra. The co-operation with suppliers of these idiosyncratic inputs, however, involves turbulence, ambiguity, and a "redistribution of improvisation rights"' that is more in tune with the fluid improvisations of jazz. Consequently, such networked production structures have distinctive organisational characteristics. The first is provocative competence, improvising and interrupting habits and routines. Secondly, they celebrate error as a source of learning. Thirdly, they are distributed systems, with tasks spread around between people. Fourthly, people alternate between roles of leader and follower, between performing as problem solvers and supporting problem solvers. Finally, they are characterised by 'hanging out', or learning-by-watching while apparently idle and not working.

There are, however, two important qualifications that must be made about networked organisational forms. First, there is uncertainty about their extent, with partial and variable empirical evidence (Gertler and Di Giovanna, 1997). Secondly, relationships between network structure and spatial form are indeterminate. Co-location or spatial propinquity may (but does not necessarily) facilitate organisational proximity by increasing the probabilities of inter-agent encounter. Conversely, organisational proximity does not necessarily require spatial propinquity. Moreover, different network relations can take different spatial forms – territorially integrated or disintegrated – while the *same* network relationships can take different spatial forms in different time/spaces (Leborgne and Lipietz, 1991, 38–9). While spatial proximity may facilitate incremental changes to process technologies, radical process innovations and the creation of new production practices require close interaction between producers and users of technologies (Von Hippel, 1998), which is most effective in circumstances in which they share triple proximity – culturally, organisationally and spatially (Lundvall and Johnson, 1994). In other cases, spatial proximity results because production depends upon material exchanges between co-located processes and firms. Outputs from one become inputs to others, notably in continuous flow industries such as bulk chemicals and steel (Hudson, 1983, 1994c). There are, however, no rigid deterministic relationships between a particular form of inter-company relations and a particular geography of the production system. In networked HVFP systems component supplier companies can be co-located or can be – literally – located on the other side of the world. Sometimes companies have established one or two factories to serve the entire global market (Howells, 1993, 227). Clearly 'just-in-time' does not necessarily equate with co-location in a shared, bounded space.

7.3.2.3 RESHAPING PRODUCTION SYSTEMS VIA LONGER-TERM STRATEGIC COLLABORATION

Strategic co-operative relationships can take a variety of forms. Some – for example, trade associations and cartels, through which firms seek to control markets and avoid 'ruinous competition' – have a long history. Others are more recent, emerging around issues of shared strategic interest, underpinned to varying degrees by 'trust' (Lorange and Roos, 1993).[8] Such alliances are forged for a variety of specific purposes. They include:

- market access and penetration – for example, gaining access to mature national markets in sectors such as ICT (Cooke and Wells, 1991) or reducing the risk of penetrating high-risk markets, such as those of the transition economies of China and central and eastern Europe (Smith, 1998), sometimes via specific joint ventures;
- coping with rapid market evolution or creating such market turbulence as a competitive strategy;
- sharing the increasing costs, uncertainties and risks of R&D and new joint product development, especially in technologically-sophisticated sectors of production, and sharing technologies (Chesnais, 1993, 19). Often this involves joint ventures specifically for this purpose;
- identifying more and less commercially promising new product areas by sharing R&D costs (Walsh and Galimberti, 1993, 187–8);
- joint production to realise economies of scale and scope (Lie and Santucci, 1993, 116) or to cope with problems of over-capacity;
- taking advantage of coalescing product categories, especially in sectors characterised by rapid technological change and heavy reliance upon expensive technologies (Hagedoon, 1993; Mulgan, 1991).

While not new, strategic collaborative ventures across national boundaries have greatly increased in number (Lester, 1998) and changed qualitatively in significance. Three features of this recent expansion are particularly striking. First, they have become central rather than peripheral to the competitive strategies of 'virtually all large (and many smaller) corporations' (Dicken et al., 1994, 32).[9] However, such alliances are most common between large transnationals, especially in sectors associated with new, technologically-intensive products, 'typified by high entry costs, globalization, scale economies, rapidly changing technologies and/or substantial operating risks' (Dicken, 1998, 228, citing Morris and Hergert, 1987). Consequently, they have a distinctive geography, concentrated within and between the global triad of macro-regions in North America, Europe and Japan. Secondly, the vast majority are between competitors. Thirdly, many companies are forming networks of alliances rather than involving themselves in a single alliance: 'relationships are increasingly polygamous rather than monogamous' (Dicken, 1998, 228). Consequently, boundaries between firms have become more blurred as the socio-spatial anatomy of production systems has become more complex.

7.4   Projects, virtual firms and new spaces of production

There has been growing emphasis on projects and temporary project teams as a new organisational model, especially within flexibly networked firms and particular sectors (Grabher, 2002). Major corporations are internally decentralised as networks, which connect themselves as specific business projects, and switch to other networks as soon as the project is finished (Castells, 1996, 151–200). Projects constitute 'temporary social systems' in which people with diverse professional and organisational backgrounds work together to accomplish a complex task, insulated from the day-to-day activities and organisational routines of the firm: 'The arch-typical action unit becomes the multi-skilled, multi-knowledgeable and temporary project team' (Hagstrom and Hedlund, 1998, 180). Such temporary project structures can also extend across firm boundaries as 'virtual firms' are created for a particular period of time to complete discrete tasks. Consequently, major corporations work in a series of changing alliances and partnerships, specific to a given product, process, time and space.

As collaborations between diversely skilled people, projects provide 'trading zones' between different business models, identities and philosophies (Grabher, 2001, 361). As such, projects and 'virtual firms' may simultaneously contain and combine different extant approaches to production, creating (temporary) hybrid forms. Moreover, their emergence may signal a significant change in corporate organisation. Assembling members of existing companies into teams to execute specific tasks is effectively a process of 'mass customisation of the enterprise', with the ability routinely to form virtual companies. This new, project-based organisational form renders the boundaries of the firm, in the conventional sense, virtually irrelevant. The key element becomes the networks of linkages within and across these boundaries. On the other hand, the emergence of project-based teams presumes that firms are identifiable legal and social entities, possessing defined boundaries and a degree of permanence and coherence, able to form contracts and co-operate on the basis of mutual interest and trust. As such, people can identify with them.

However, the formation of 'virtual firms' or 'project-based teams' cannot escape underlying contradictions in the social relations of capital. The socialisation of knowledge production makes it hard to distinguish legally between the intellectual property of different firms. This reinforces a tendency for network economies to be captured by the network and, in turn, for new forms of enterprise to emerge that are able to capture such economies without destroying the broader network that generates them. 'Virtual firms' are said to correspond to this need. Some companies already approximate such virtual enterprises (Pine, 1993, 258–63). However,

> unless the virtual firm becomes co-extensive with all those involved in production, the contradiction [between the increasing socialisation of the productive forces and the private control of the social relations of production] is still reproduced on the side of the social relations of production. For whereas every capital wants free access to information, knowledge and expertise, it also wants to charge for the information, knowledge and expertise that it itself can supply. (Jessop, 2000, 5)

This is a salutary reminder that while variation in organisational form is important, so too are the structural parameters of a capitalist economy.

7.5   Spaces of knowledge production and learning: firms, networks or territories?

There has been a shift in emphasis from specifically designated spaces of knowledge production, such as R&D labs or design studios, to recognition that everywhere in (and often beyond) the firm constitutes a potential site of knowledge creation, expressed forcibly in the concept of *ba* as a fluid and relational space of knowledge creation. Lundvall (1992) stresses the significance of national innovation systems, of shared language and culture, as well as formal legislative frameworks, in shaping trajectories of innovation and learning. Others privilege the local or regional over the national in the production of knowledge and learning, heavily influenced by the 're-discovery' of communities of producers in industrial districts in the Third Italy. Similar manufacturing districts have been identified in several European countries (Garofoli, 2002), as well as in advertising and financial services in London (Amin and Thrift, 1992; Grabher, 2001). Such districts possess an institutional capacity continuously to learn, adjust and improve in economic performance, based on intersecting non-hierarchical networks with multiple apexes in which no single firm is permanently at the head of any hierarchy. Complex networks of territorially agglomerated SMEs are seen as the main – maybe the only – route to learning (Best, 1990, 234–7) although Garonna (1998, 228) disputes this, arguing that learning in industrial districts involves 'virtuous circles' of relations between SMEs and large companies. Others seek to develop ideas of regionally-based learning without necessarily privileging networks of SMEs (for example, Braczisch et al., 1998; Morgan, 1995).

Linked to this, there has been growing emphasis upon non-economic relationships, territorially embedded shared values, meanings and understandings, tacit knowledge and the institutional structures through which it is produced, in underpinning regional economic success and regionally-based learning systems. Focusing on the region as a nexus of untraded interdependencies (Storper, 1995) decisively shifts the emphasis from firm to territory as the key institutional form and space for creating and disseminating knowledge and to a collective, culturally and socially embedded, competitive entity. Moreover, as codified knowledge is increasingly ubiquitously available, tacit knowledge, rooted in relations of proximity, is increasingly regarded as the key source of competitive advantage.

There is, however, no a priori reason to privilege territorial over corporate knowledge production and learning (or vice versa), or to privilege any particular spatial scale or size of firm, irrespective of context. Distributed innovation systems can involve complex arrangements of inter-firm alliances and relationships between spaces and scales of knowledge production.

Regional and locality-based learning and knowledge production systems can be significant (Maskell et al., 1998), especially if innovation systems are constituted sectorally – and at least potentially globally – rather than nationally (Metcalfe, 1995). The sectoral constitution of innovation systems across national boundaries emphasises the significance of local specificity within global production systems and spaces and of links between corporate learning and territorially embedded knowledge. There are also powerful pressures on companies to collaborate via international alliances and joint ventures to create knowledge and access scientific and technological advances made in foreign countries. As a result, 'external sourcing of knowledge, and its cross-border exploitation and application, are growing rapidly, with signs of increasing national specialisation by sector' (Radice, 2000, 734).

Competitive success depends upon the distinctive ways in which companies produce and use different types of knowledge. For example, although 'poorly shared, ambiguous meanings that work through affect rather than signification may defy explicit coding', they are nevertheless 'recognised, reacted to and judged by the cues available as part of something creative and knowledgeable' (Allen, 2002, 55). In addition, firms must mobilise codified and tacit knowledges for competitive advantage, blending action at a distance and local practices within organisational spaces (Amin, 2000). Globalised forms of organisation are predicated upon disembedding fragmented products of local learning from the spaces in which they were initially produced, and integrating them to serve strategic corporate interests. Success in complex environments depends upon governance cultures that help generate variety and mixtures of competences rather than privileging of one type of knowledge over another.

Major firms are developing dual governance and organisational structures, dealing with qualitatively different types of function (core and non-core) and requiring qualitatively different types of knowledge and learning mechanisms (knowledge processing and information dissemination). The key issue for the firm is successfully coupling the two sets of mechanisms and securing their successful reproduction in the day-to-day practices and transactions of the firm, its workers, and the communities of practice that they form within and across its boundaries. Such communities, groups of people informally bound together by shared expertise and a common problem that they seek to solve by collective learning–in–working, represent the most effective mechanism through which tacit knowledge is created and diffused within companies (Gertler, 2001, 18). The potentially disruptive effects of multiple views and polyvocal philosophies becoming noise within heterarchic firms are held in check by 'tags', which define rules and protocols for shared understanding (Grabher, 2001, 354). Moreover, the effective reproduction of communities of practice and knowledge creation can be facilitated 'through the construction of office spaces which can promote creativity through carefully designed patterns of circulation ... [based on] integrative spaces generating transactional knowledge via interactive group work and connected team projects on an as-need basis' (Thrift, 2002, 224). For the 'challenge posed by the notion of heterarchy is adaptability' (Grabher, 2001, 354), that is, learning to learn, 'to *evolve in order to adapt*' (Amin and Cohendet, 1997, 6, emphasis in original), and the creation of spaces to enable and facilitate this.

7.6 Redefining corporate anatomy and spaces of production via acquisition and merger

Companies can alter the anatomy of ownership and control of production via (amicable) merger and (sometimes hostile) acquisition (M&A) activity. As well as variability by industry and sector, M&A activity displays marked temporal and spatial variability. Corporate amalgamation involves major social restructuring and so 'is bound to run into roadblocks. The result is a wave-like pattern, with long periods of acceleration followed by shorter down-turns' (Nitzan, 2001, 234). This periodicity is also related to changing state regulatory regimes and spaces. Such frameworks may prevent or discourage mergers because they would pose a threat to competition or encourage them to produce national or supra-national 'champions'. Similarly, and related, some cultures of capitalism encourage M&A (notably the Anglo-American model), while others are or until recently have been less receptive to such corporate behaviour, especially if this involves unwelcome hostile acquisitions (the German and Japanese models, for example).

However, it was not until the 1970s that processes of M&A became global rather than national, linked to changes that partly denationalised spaces of regulation. For example, the considerable increase in M&A activity in the 1980s and 1990s was linked to global deregulation of financial markets, the creation of 'macro-regions' such as the Single European Market and the North American Free Trade Area, and growing competitive pressures on companies in increasingly internationalised markets. Most Foreign Direct Investment (FDI) in the 1980s was in the form of M&A, a change directly linked to brownfield replacing greenfield as the dominant form of FDI (Nitzan, 2001, 251–2), with a majority of investment in services (Weiss, 1997), including financial services (Martin, 1994), producer services (Coe, 1997) and advertising (Leslie, 1994). The development of transnational M&A is linked to financial globalisation, the formation of 'totally internationalised' financial and monetary markets, enabled by the growth of ICTs, and international trade agreements (Chesnais, 1993, 13). This combination of factors laid the foundations for 'world oligopoly' to become the dominant form of supply structure in R&D intensive and 'high-tech' industries characterised by rapid technological change as well as in 'scale-intensive' manufacturing industries. Increasingly, open international markets expanded the scope for intra-industry expansion across national boundaries while legitimating further national concentration in pursuit of 'global competitiveness'. However, 'such refocussing is bound to become exhausted, pushing dominant capital towards renewed conglomeration on a global scale' (Nitzan, 2001, 246). However, as Chesnais hints, the extent to which M&A led to marked centralisation of capital and the emergence of truly international – let alone global – firms is uneven, sectorally and spatially. While evident in sectors such as computing, communications, chemicals and pharmaceuticals, in many others the supply structure remains national, or sub-national.

Companies engage in M&A activity for a variety of reasons (Waters and Corrigan, 1998) and the same company may pursue several of these simultaneously:[10]

- Big companies acquire innovative small companies to access growth products or process technologies, control the pace of their diffusion, or broaden their product range or acquire technological rents, while avoiding the R&D costs of developing these (Nitzan, 2001, 253). Conversely, market structures may pressurise small companies to sell out, as in the high-tech electronics sector industrial districts of late twentieth-century California (Storper, 1993, 447–8). Even in the canonical 'horizontally' networked industrial districts of the Third Italy, there have been growing tendencies to sell out, buy up and vertically integrate production (Hudson, 2003).
- M&A between major companies in R&D intensive activities, such as automobiles, electronics and pharmaceuticals, allows the spreading of costs (Pilling, 2000).
- M&A allows penetration of new markets via acquiring ready-made distribution networks. Market conditions of stagnant or slowly growing demand encourage the growth of M&A (Chesnais, 1993, 17).
- Acquisition, followed by rationalisation and plant closure, allows companies to eliminate surplus capacity. However, the potential impacts of employment loss can deter M&A, even in sectors characterised by considerable excess capacity (such as the automobiles in the European Union: Hudson, 2002).
- Companies acquire other companies to create synergies via complementarity in product ranges.
- SMEs may merge in response to growing competition from big companies and smaller more specialised niche producers: for example, in advertising and marketing (Nixon, 2002, 136).
- Companies acquire others to diversify and broaden their product range, and/or change it via moving into some and out of other product areas. During the 1990s, for example, ICI engaged in a series of selective acquisitions, asset swaps and disposals of peripheral and/or weaker activities to transform it from a bulk commodity to a speciality chemicals producer (Edgecliffe-Johnson, 1998).
- Acquisition can be a way of getting bigger, increasing market share and volume of output, classic processes of concentration of production and centralisation of capital. For example, this occurred in the UK in the 1980s in clothing (Crewe and Davenport, 1992) and in the cleaning, catering and security industries (Allen and Henry, 1997), in the 1990s in advertising (Grabher, 2001) and pharmaceuticals (Pilling, 2000) and in a range of global mining companies (O'Connor, 2000).

- Companies merge to control markets, although seeking to oligopolise a sector may be problematic, as in the USA in the late nineteenth and early twentieth centuries.
- M&A allows companies to spread risks by diversifying into products with varying business cycles, or into different spatial markets for the same products.
- Finally, companies may acquire other companies simply because of 'fashion effect', a perceived need to increase in size and not to be left behind in the face of M&A activity by rivals.

Despite the rhetoric about the revival of SMEs, 'big' is still 'beautiful' in capitalist economies. Production in many sectors and products continues to become concentrated in a few giant firms via processes M&A and centralisation of capital. Size confers many competitive advantages. These are not, however, unconditional or automatically conferred advantages. Problems of 'diseconomies of scale' can result from M&A creating 'conglomerates which are considered as notoriously ineffective forms of organisation' (Ramsay, 1992, 31). Consequently, disposal often follows shortly after M&A as companies seek to maximise the benefits without carrying the costs of increasing size.[11]

7.7   The Internet, cyberspace and spaces of e-commerce: new models of intra- and inter-company organisation

Much has been claimed about the alleged transformatory effects of the growth of the Internet on the economy as 'materiality becomes subordinated to information and knowledge creation' (Kenney and Curry, 2001, 48). The Internet certainly represents the latest advance in technology for transmitting codified information, reduced to its most abstract form of 1s and 0s, a giant machine for reducing transaction costs, or in Bill Gates' terms, a tool for friction-free capitalism. As the real cost of processing power, bandwidth and connections continues to fall, 'it is reasonable to assume that anything that can be digitised will be'. In particular, business-to-business (B2B) transactions[12] have been at 'the leading edge of the e-commerce revolution' and 'electronic commerce among businesses is expanding in dramatic fashion as firms realise that in order to compete effectively they must interact and carry out business using variety of electronic means' (Leinbach, 2001, 17). Some 80% of all e-commerce transactions in the USA in 1999 were B2B (Button and Taylor, 2001).

By the late 1990s advertising companies such as WPP were spinning off new subsidiaries to broaden and deal with their Internet operations, while taking minority shares in companies located in the 'highly promising and equally risky terrain of interactive marketing services'. However, exemplifying the rapid pace of Internet developments, 'the drift towards new media, triggered by the accelerating pace of spinning off, merging, and starting up of new interactive businesses, in 1999 culminated in the foundation of Wpp.com, an internal holding company of all web-based activities of WPP' (Grabher, 2001, 358–9). There was also a proliferation of websites seeking to consolidate B2B markets around particular product segments (such as food and packaging). In other cases, major manufacturing companies established their own online exchanges. For example, GM, Ford and Daimler-Chrysler established a unified exchange for automobile components. They and their first-tier suppliers use the site to purchase from second-tier suppliers. The site is projected to process some $240bn of transactions annually, cut purchase order processing costs from $100 to $10, reduce inventory holdings, facilitate mass-customised production, and enable faster delivery.[13]

B2B largely involves using a new technology within existing business models, allowing more sophisticated supply chain management, with greater cost and quality control in exchanges of material commodities between companies and is stimulating further development of HVFP. Routine, standardised activities and transactions that have a separable information content are especially amenable to digitisation and transmission via the Internet. B2B replaces people occupying 'the mechanical segments of the information exchange pipeline' with online transactions, and 'allows every step in the procurement process [to] be monitored and optimised'. This has led to significant estimated B2B transaction cost savings: 39% for electronic components and 20% for computers (Button and Taylor, 2001, 33). However, B2B markets are social constructions, involving power relations between those that create and participate in them. This is especially so in the B2B arena 'as the owner of the transaction platform has the potential to control the transaction conducted, both in terms of the rules but also in terms of rents'. For example, the control of the automobile site by the 'big three' assemblers and their resultant access to information about suppliers' behaviour and cost profiles will further reinforce their power within the supply chain (Kenney and Curry, 2001, 62–3).

The speed with which B2B businesses can be established places a premium of 'first mover advantage', being the first in a particular market category (although the history of e-commerce to date emphatically demonstrates that being first is no guarantee of commercial success). For example, the creation of the 'big three' automobile supplies website dramatically reduced the scope for competing sites to be established and 'rather than permit interlopers to capture the benefits from becoming electronic intermediaries, the market leaders can capture the benefits of their sponsored start-up'. As such, the governance of a B2B market is very important and 'there are pitfalls that discourage entry'. Development becomes path-dependent 'because once an exchange becomes dominant, all the users incur large switching costs that block participants from exiting' (Kenney and Curry, 2001, 62–3).

Some see the Internet as more than just a machine for cutting transaction costs. For example, the Internet 'is a newly developed space with the power to give rise to novel forms of human social interaction' and so it is 'highly likely that new processes will emerge from the [Internet] itself. This is concretised in new ways of using the Internet or when new e-commerce models are "invented" or "discovered" and implemented in code (a new software programme)' (Kenney and Curry, 2001, 46). Thus, it is claimed that e-commerce is giving rise to new business models and practices, and that it will continue to do so:

As long as the Internet remains an essentially open platform, its ability to develop novel approaches will likely remain high. The most successful enterprises of the future will be based on the Internet's own paradigms rather than paradigms based on the past which [will be] rendered meaningless by the collective imagination and creativity of cyber explorers who are only beginning to learn the true contours of the new world they have created. (Kenney and Curry, 2001, 64)

However, assertions as to knowledge creation and the emergent socio-technical properties of the Internet, and its capacity to effect radical economic transformations and become the basis of new forms and models of capitalism, are contentious and conjectural. Whether such bold claims are justified remains, at least as yet, an open question and Kenney and Curry recognise that 'the final configuration caused by the Internet is difficult to [predict because the features of the Internet interact in problematic and contradictory ways'. What is certain is that as long as cyberspace is produced within the contradictory social relations of capital, it will have to respect the structural constraints, value relations and material realities of *that* paradigm and this will shape its development.

7.8   Spaces of production beyond the formal economy

There is a variety of spaces of production within capitalist economies, of varying degrees of (in)formality and (il)legality, which, taken together, are of some significance. Indeed, the 'informal economy' (that is, activities that are legal but performed beyond the 'normal' regulatory regime) is 'not a marginal but a fundamental political economic process at the core of many societies' (Castells and Portes, 1989, 15). In other parts of the world the dominant political economic processes relate more to activities that are illegal (for example, historically in areas such the Italian Mezzogiorno (Arlacchi, 1983) and contemporarily in much of Russia (Dunford, 2004).[14]

The translation of employed labour from the unsupervised space of the home to the controlled and surveilled space of the factory was central to the emergence of industrial capitalism. Homeworking for a wage has, however, never been abolished and in some circumstances continues to flourish, even expand. McRobbie (2002, 99) notes the blurring of the boundaries of the workplace and 'the encroachment of work into every corner of everyday life, including flexible working at home'. Companies resort to homeworking in response to tight labour market conditions and/or to exploit female labour that is only available for work in the home residence (Hadjimichalis and Vaiou, 1996; Peck, 1995) while women accept homework in order to make domestic and waged work compatible. In countries on the periphery of global capitalism and in urban areas within more central parts of the capitalist world, extensive homeworking is typically associated with piecework in mature, low-technology and labour-intensive consumer goods industries such as clothing (Leontidou, 1993; Portes et al., 1988). However, companies operating in core areas in high-tech activities may alter their recruitment and retention strategies in the face of growing skill shortages,

FIGURE 7.2 Informal work; shining shoes in Lisbon, Portugal

introducing new systems of homeworking to allow women with children to combine child care with paid work (Summers, 1998). Advances in ICTs have also enabled an increased number of professionals (such as accountants and designers) to work from home. Overall, however, the expansion of homeworking as a result of advances in ICTs has been modest. While easing recruitment problems and lowering fixed capital investment costs, homeworking also exacerbates problems of control precisely because of the spatial separation of workers and managers. The spatial configuration of spaces of production between home, factory and office reflects a trade-off between the need to recruit suitable labour and the weakening of control of the labour force that may accompany homeworking.

In other activities, such as agriculture and construction, in which work of necessity is carried out in much more open, permeable spaces, less amenable to close supervision, surveillance and regulation, there is a long history of informal work, shading into illegal work – for example, when the work is legal but those employed to do it are illegal migrants, or when the things produced themselves are illegal, such as drugs. Equally, in manufacturing activities such as clothing sweatshops in major cities and urban areas, work is often performed by illegal migrants or by young people below the minimum age of legal labour market participation. Such people are vulnerable to intensive exploitation, precisely because of their lack of legal status and/or the rights of formal economic citizenship.

For many years, the informal sector was seen as a characteristic of economies on the peripheries of global capitalism, especially the labour markets of their urban areas, as migrants from the countryside were unable to find formal sector employment but needed to find way of 'getting by', of devising personal and household survival strategies (Redclift and Mingione, 1986). More recently, such activities have again expanded in the core territories of capitalism, especially their major urban areas, but also in areas that have experienced severe industrial decline (Williams and Windebank, 1998). The collapse of state socialism led to a dramatic rise in those employed in the 'irregular' economy over much of central and eastern Europe, Russia and the Confederation of Independent States (CIS) as the jobs, wage incomes and the welfare safety net of the previous era were dramatically eroded in the transition to capitalism. As a result, many people have had no choice but to explore possibilities beyond the formal economy, precisely because the latter has little to offer them (Dunford, 2004).

Finally, in many parts of the world, in the interstices of the mainstream economy or in its fringes, there is a variety of spaces of production linked to different conceptions of the economy from that of the capitalist mainstream, variously described as the voluntary sector, the social economy, the 'Third Sector' and so on. Fundamentally, these are spaces in which production, both of material goods and services, is informed by a variety of non-profit motives, such as a concern for human welfare, equity, social and environmental justice and so on. Paradoxically, national states and the European Union have shown a growing interest in promoting such alternative spaces via public policies, precisely because of the ineffectiveness of the mainstream in tackling developmental problems in marginalised cities and regions. Such spaces thereby come to occupy an important legitimating role, but at the same time hold open the conceptual and political possibilities of alternatives to mainstream orthodoxies and spaces (Leyshon et al., 2003).

7.9   Summary and conclusions

Within capitalist economies, there is a variety of spaces of production and of the governance and social relationships through which production within them is organised and made possible. This socio-spatial anatomy of production has always changed and evolved over time, and is expressive of the restless character of capital as it searches for profits in new spaces, products and people. The growing complexity of production is registered both in new forms of organisation within firms and in new forms of organisation which involve inter-firm relationships but also, in some respects (as in project working and various forms of 'virtual' organisation), render the boundaries of firms problematic and of diminished relevance.

While the organisation of production in the mainstream economy has evolved, this has created a complex mosaic of forms and spaces of production, linked in often complicated ways. As new methods of production have been created, and new spaces of production constituted, these have become overlain on and intertwined with the existing socio-spatial anatomy of production. Consequently, rather than a linear sequence of one model of organisation replacing the previous one, eradicating its spaces and constructing new ones, different models and spaces of production have become interwoven, so that production is always in the process of becoming. Furthermore, alongside, and to a degree woven into production in the formal mainstream economy and its spaces, diverse forms of informal and illegal production in 'shadow economies' have continued to flourish, within the interstices and on the fringes of the mainstream. Thus production in capitalist economies involves a range of social relations and spaces beyond those of the formally regulated mainstream.

Notes

1 While I distinguish between the organisation of management (section 7.3.1) and the organisation of work by managers at the point of production, this dichotomy fails to capture important variations in power, autonomy and control among 'managers' and 'workers'.

2 In value theoretic terms, surplus-value production can be enhanced by increasing either absolute or relative surplus-value (Aglietta, 1979, 49–52). More prosaically, it involves lengthening the working day, intensifying the pace of work, or increasing labour productivity by technological innovation.

3 The latest generation of integrated circuit plants has the capacity to produce 5% of global output, with higher productivity and lower costs than their predecessors (Foremski et al., 2003).

4 HVFP also has implications for work organisation in retailing (see Chapter 8).

5 Such personal satisfaction is also important in interactive service work (see Chapter 9).

6 See section 4.2 and also R. Hudson, 2001, Chapters 4 and 5.

7 The commodity chain analysis distinction between 'producer-driven' and 'consumer-driven' chains disguises the variety and subtlety of network relations and governance structures and their spatialities (Smith et al., 2002).

8 For example, in the USA the percentage of turnover of the largest 1,000 companies generated through strategic alliances doubled between the early 1990s and 1998 to 25% (cited by Larsen, see www.iri.com).

9 Regulatory policies and practices of national (and emergent supra-national) states may encourage or deter such forms of co-operation (Mulgan, 1991).

10 These often mirror reasons for forming strategic alliances. Indeed, firms choose between M&A and strategic alliances, depending on which best meets their objectives (Thompson, 1999).

11 However, cross-national mergers can pose problems for national states (see Chapter 6).

12 Business-to-consumer – B2C – and consumer-to-consumer – C2C – models are discussed in Chapter 8.

13 Initially each company planned to build its own website but separate billing systems would have created duplication and difficulties for suppliers.

14 It has been estimated that the 'shadow economies' are equivalent to 8–30% of GDP in developed economies, 7–43% in transition economies, and 13–76% in developing economies. Allowing for definitional variation and problems of non-availability of data, even the lower limits indicate the aggregate importance of these shadow economies (see www, ilo.org, accessed 01/04/2003).

8 Spaces of Sale

8.1 Introduction

Understanding the economy requires serious consideration of exchange, sale and consumption. The exchange of money for commodities in the formal economy, performed in dedicated spaces of sale, is a critical moment in the realisation or surplus-value and the process of capital accumulation and a prelude to subsequent (final) consumption. Spaces and practices of consumption, circulation and exchange are central to a reconstructed economic geography (Crang, 1997). However, while clearly linked, exchange and sale are analytically different from consumption. It is important not to conflate them, although purchasing commodities may involve consumption of spaces of sale. Such spaces are simultaneously material sites for commodity exchange and symbolic and metaphoric territories, 'contested sites where the identities of individuals and commodities are given meaning' (Lowe and Wrigley, 1996, 20). More generally, social integration 'takes place through the seduction of the market place, through the mix of feeling and emotions generated by seeing, holding, hearing, testing, smelling, and moving through the extraordinary array of goods and services, places and environments that characterise contemporary consumerism' (Lash and Urry, 1994, 296). This is a strong claim, one that must be carefully circumscribed in terms of its validity in time/space within capitalism, for exchange is governed by different logics in other non-capitalist and non-mainstream circuits.

The construction of major retailing and commercial centres is linked to the emergence of secondary circuits of capital, developed to absorb surplus-value that could not be absorbed in the primary circuits. The evolution of spaces of sale has been linked to the development of different methods of production, in particular the rise of mass production and the subsequent evolution of post-mass production approaches. It has also been connected to changes in the anatomy of retailing, particularly via processes of merger and acquisition. For example, there were pressures to reconfigure spaces of sale in North America and western Europe from the 1980s as a small number of companies came to dominate retailing, especially in food and clothing (Crewe, 2000, 276). For a variety of reasons, therefore, such investments in the built environment take varied spatial forms, constituting diverse spaces of sale. These range from the grand and spectacular – the mega-mall or giant superstore – to the more mundane spaces of the corner shop. There is an identifiable historical-geography to their evolution – 'an arc from the arcades and department stores of Paris through to the shopping malls of the United States' (Miller et al., 1998, 3). Such formal spaces and their associated organisational forms (for example, the supermarket giant) and exemplary firms increasingly are diffusing internationally (for example, into central and eastern Europe and China) as new spaces open up to retailing capital.[1]

While 'the shop' or 'the store' (and one could add, the mall) often forms the centre point of geographies of retailing, they form only one of many channels through which goods might be bought. Spaces of sale (or non-monetary exchange) may be differently constituted, in part depending upon relationships between systems of production, exchange and consumption (for example, as the nodal points and routes of peripatetic markets: Berry, 1967). Nonetheless, 'what is remarkable is how far back in time the history of retail shops and shopping stretches; the scale on which shops and shopping operate; and how much of this history ... is based on a reflexive relationship between consumers, shopkeepers, and sites of consumption (understood as streets, markets, shops, galleries, and so on), sites which act as an active context rather than a passive backdrop' (Glennie and Thrift, 1996a, 26). These sites range from diverse fixed locations (the West Edmonton Mall, corner shops, peoples' residences), to the mobile spaces of car boot sales and jumble sales, and the informal/illegal markets of streets and street corners, continuing the pre-capitalist legacy of peripatetic spaces of sale in new forms.

8.2   The formal spaces of the shop and department store

The great department stores of the nineteenth century (predominantly in London, New York and Paris), and their twentieth-century successors, have been characterised as 'one of the classic consumption spaces' (Wrigley and Lowe, 2002, 203), 'the quintessential consumption site of the late nineteenth and early twentieth century' (Jackson and Thrift, 1995, 18), 'palaces of consumption ... the most visible, urban representations of consumer culture and the economics of mass production and selling' (Domosh, 1996, 257). While such stores can and do function as consumption spaces, they are first and foremost 'classic and quintessential spaces of sale'; their rationale rests in the sale of commodities and realisation of the surplus-value that they embody. Such stores have often been explicitly designed to promote consumption as a way of increasing sales, however. In so far as they do develop as consumption spaces, this is primarily as a corollary of, and route towards, their primary function of selling things. However, such grand stores only constitute a small minority of shops and stores, compared to the much larger number of more mundane, 'ordinary' shops.

Sales in these spaces – whether in grand stores or corner shops – are typically conducted on the basis of fixed prices for a given standardised good, non-negotiable between salesperson and purchaser, with legal and regulatory frameworks that formally define the rights and responsibilities of buyers and sellers. The act of purchase involves comparing the relative worth of different items, assessing quality and desired attributes against a given price. Shopping became a skilled, knowledge-based activity, albeit with that knowledge heavily influenced by retailers and advertisers. Department stores provided spaces in which women could be taught to purchase, and to some extent how to consume, to regard encounters with a range of consumer goods as a norm of everyday life. This was achieved by design and control of gendered spaces, the use of advertising and the display of commodities as spectacle and the use of demonstrations and specialist shop staff. As such, the shop and store both enable and constrain the ways in which sales can occur.

Furthermore, the shop or store also provides an opportunity for retailers to distinguish themselves and their commodities (Dowling, 1993). Creating distinctiveness may involve working on the exterior and/or interior appearance and design of the store. Retailers can use design as a strategic business resource in one or more of four ways: to differentiate; to focus or segment operations; to re-position stores; and to represent stores as brands, fixing their image. For example, the strategy of the Next chain of clothing stores centred on the presentation of a limited and co-ordinated garment collection, targeted to specific groups of consumers. More specifically (and initially) 'young working women who were weary of the fast fashion in the High Street boutiques but not weary enough for the staid styles of the Department stores'. The men's wear range was marketed 'effectively to *address and shape* an "upmarket" but affordable middle market in menswear'. Next thus identified 'underserviced' segments of the clothing market and its retailing strategies focus on effectively servicing those segments (Nixon, 1996, 50–2, emphasis added). Design also involves manipulating space within the retail environment, using spatial strategies within stores to boost sales. As goods became increasingly standardised, the spaces of sale in which they were acquired and the activity of shopping assumed increased significance.

In summary, the great department stores, constructed discursively, socially and materially in the late nineteenth and early twentieth centuries in the major cities of North America and western Europe, constitute one of the 'key iconic aspects of modern urban society' (Nava, 1997, 56). For example, the new department stores of New York were much larger and differed from their predecessors 'in the degree of ornamentation, the attention to detail and display of goods, the concern with the internal organization of the departments, and the catering to the personal needs of the shoppers, most of whom were women' (Domosh, 1996, 264). The predominance of female shoppers was not accidental. Retailers targeted women as their customers and deliberately cultivated associations between women, fashion and religion. At one level, this required learning to schedule the 'openings' of collections to coincide with Christmas and Easter, further aligning religion and fashion, and institutionalising the commercialisation of religious holidays and associated gift giving. Department store owners further elaborated the association between women, fashion and religion by calling their stores 'cathedrals' and their goods 'objects of devotion'. Nava's (1997) comparison between Selfridge's and Westminster Abbey exemplifies the way in which these new department stores were expressions of a new, almost religious, fervour of purchase and consumption. With stores as 'cathedrals' and women as 'worshippers', purchasing became a moral act, a religious duty. However, in order to preserve legitimacy and maintain appropriate gender roles these sacred spaces of sale had to be feminised, 'to appear as cultural and civic spaces not completely tainted by commercialisation'. As spaces of sale dedicated to promoting mass consumption, there were limits to the ability to banish all traces of commercialisation. At the same time, they had to become a safe public arena, insulated from the chaos and dangers of the street, in which (respectable, middle-class) women could safely shop. These spaces of sale needed to become 'an appropriate feminine environment. The qualities associated with nineteenth-century femininity and the domestic sphere were built into the store: there were symbols of civic and cultural aspirations, well-ordered and arranged displays, services and amenities designed for women, and an environment meant to be safe and protected' (Domosh, 1996, 265–70).

Another central theme of these new retailing spaces was their role as centres of entertainment and tourist attractions – long before the late twentieth-century post-modern concern with 'de-differentiation' entered the vocabulary of contemporary social science. Such department stores rapidly became focal points for social life in the urban environment. Nevertheless, they were primarily spaces of sale, driven by hard capitalist logic. As such, increasing sales was never far from the agenda, unavoidably given the decisive relationships between mass production and successful sale of consumer goods. For example, providing varied services such as banks, exhibitions, and restaurants helped attract potential customers living some distance from the store. Furthermore, as many of Selfridge's adverts insistently reiterated, its prices were 'the lowest – always'. By introducing from the outset the American innovation of sales and bargain basements, Selfridge helped expand the class spectrum of its targeted customers. Put another way, there was recognition that these were class-divided spaces of sale and that boosting sales volumes entailed blurring those class divisions. The new spaces of religion/sale, the new cathedrals to consumption, had to be democratised, made available to a wide social spectrum. In so doing, they soon developed as everyday arenas in which a growing spectrum of women could safely engage, a private space that became part of a public sphere (Nava, 1997).

The great department stores are only one form of store. During the middle decades of the twentieth century, more mundane forms of shop were increasingly reorganised as spaces of sale, with the specific aim of redefining purchasing behaviour in relation to particular categories of commodities and gendered purchasers, increasing sales and profits within modern society. Often this involved moves from counter service by skilled retailing staff to self-service. For example, the modern mode of shopping for food in Woodward's department store in Vancouver was primarily scientific, rational and new. The food floor became self-service after the 1940s. Aisles were wide, commodities well organised, both facilitating a modern mode of shopping that consisted of methodically searching for desired commodities. As such, shopping now encompassed a rather different form of work for the shopper and 'food floor shoppers had to be taught how to be modern'. As well as a space of sale, Woodward's food floor constituted a site for learning-by-doing as women would there learn how to cook, what new and modern foods were available and hence 'how to be a better wife and mother' (Dowling, 1993, 314). In contrast, elsewhere spaces of sale were associated with definitions of masculinity rather than femininity. For example, the UK clothes retailer Burton assiduously sought to construct its shops as male spaces of sales, stripped of any association of 'feminine culture' (Mort, 1996, 138). Thus 'modernisation' of spaces of sale could be, and typically was, a sharply gendered process.

More recently, a rather different type of store has sprung to prominence within the major cities of the late modern world, such as London and New York, again drawing on the social function of the store as a meeting place and a site of interaction and entertainment as well as a space of sale. These 'flagship' stores, incorporating their 'own label' as their sole product identity, especially when clustered into 'streets of style' (discussed more fully below) typically focus upon specific groups of consumers and lifestyles. Building on concepts developed by chains such as Habitat, new designer stores sell 'lifestyle' rather than simply things and, arguably, are increasingly taking on the attributes of mini-department stores, '[combining] within their "spaces", designer interiors, rituals of display, and leisure, sexuality and food. The association of products and place identity is profound' (Lowe and Wrigley, 1996, 25). This conception of a space of sale reaches its apotheosis in the concept of 'entertainment retail'. Such prototypical stores (for instance, Nike Town) have become major landmarks in the urban landscape and 'like the old department stores, entertainment retail stores enjoy favourable coverage in local newspapers for their "enchantment" of the urban landscape' (Zukin, 1998, 883). The enchantment that results from varied non-shopping activities captures the shopper's imagination by inviting her/him to participate in simulated forms of non-shopping entertainment and experience a culture and way of life rather than simply offering the chance to buy specific commodities. Such schemes have, however, great potential to increase sales. For example, the growth of the bookshop/coffee house in North American and western Europe is the latest in a long line of elaborate techniques designed to keep the customer in the store for longer – and so more likely to buy. Even so, this way of life is only accessible to certain social groups: the 'sociality' of the store is 'dependent on visual coherence and security guards, a collective memory of commercial culture rather than either tolerance or moral solidarity' (Zukin, 1998, 834).

8.3   The formal spaces of the mall

The shopping mall is widely regarded as the iconic space of contemporary retailing, the 'urban cathedral' of fin-de-siècle late capitalism (Goss, 1993). While planned shopping centres had been constructed in the USA from the early twentieth century, shopping malls as now commonly understood appeared with the opening in 1956 of Southdale, Minneapolis, the world's first fully enclosed mall. Designed by Victor Gruen, it then 'represented innovation, creative problem-solving, and aesthetic daring, as well as shopping centre heresy' (Kowinski, 1985, 117). Southdale was the precursor for a series of developments that focused on increasing sales by manipulating the internal spatial arrangement of the mall. Crawford (1992, 13–14) summarises these as follows:

> Limited entrances, escalators placed only at one end of corridors, fountains and benches carefully positioned to entice shoppers into stores – control [their] flow ... through the numbingly repetitive corridor of shops. The orderly processes of goods alongside endless aisles continuously stimulate the desire to buy. ... The jargon used by mall management demonstrates not only their awareness of these side effects but also their ... attempts to capitalize on them. The Gruen Transfer ... designates the moment when a 'destination buyer', with a specific purchase in mind, is transformed into an impulse shopper, a crucial point immediately visible in the shift from a determined stride to an erratic and meandering gait.

Thus the behaviour of the shopper becomes a diagnostic variable in identifying the success of spatial strategies to boost sales.

The Southdale Mall formed the prototype for most malls subsequently built. Serial monotony became the order of the day as between 1960 and 1980 'the basic regional mall paradigm was perfected and systematically replicated' (Crawford, 1992, 7), with some 30,000 built in North America. To some extent such serial reproduction has spread to other parts of the contemporary capitalist world such as the UK in the form of major regional shopping centres including Brent Cross in north London, the MetroCentre at Gateshead, Meadowhall near Sheffield and Lakeside at Thurrock, east London. However, the concept further evolved in the USA with the construction of the Horton Plaza in San Diego in 1977. Designed by the Jerde Partnership, it replaced the 'dumbbell' layout with a more complex internal geometry, intended to confuse and literally lose the shopper in its multidimensional programming (Goss, 1993). This resulted in a combination of pathways of varying widths, covered and open, with staggered levels, balconies, towers, bridges, and nooks and crannies, mixing shops with entertainment and leisure facilities. These innovations encouraged 'people watching', as shopping becomes a *passegiatta*.

Moreover, there was a marked scale shift with the emergence of 'megamalls' in North America, in part in response to the proliferation of regional

FIGURE 8.1 Creating new spaces of sale: Lisbon, Portugal

malls. New super-regional malls began to be built at freeway interchanges (in turn providing stimuli for the emergence of major new suburban housing developments and edge cities: Garreau, 1991). Examples include the Galleria, near Houston, South Coast Plaza in Orange County, and Tyson's Corner near Washington DC. However, two mega-malls completed in the 1980s and 1990s – the Mall of America in Minnesota (also designed by the Jerde Partnership) and, especially, the West Edmonton Mall – exemplify the growing emphasis towards developing the mall as an entertainment, leisure and tourist attraction, a space to be consumed as well as a more effective space of sale, a way of getting people to spend more.

An important shared feature of these malls and centres (and one that is in strong contrast to many more 'traditional' shopping streets) is that these enclosed environments provide a regulated micro-climate but, perhaps more significantly, a closed space of surveillance. Combinations of CCTV and private security guards control access and record behaviour, providing – for those regarded as legitimate consumers – a sense of risk-free shopping. Customers can get on with the business of buying, with the risk of fear to personal safety greatly reduced in these privately owned and controlled 'quasi-public' spaces of sale. The West Edmonton Mall exemplifies this Foucauldian process of deliberately very visible surveillance. The hub of these activities is the electronic Panopticon of its security headquarters, central Dispatch. Within this glass-walled high-tech command post, lined with banks of closed-circuit televisions and computers, the Mall is constantly monitored by uniformed members of its security force. Patrons are very aware of its omnipresence, as routine security activities become a theatrical spectacle of reassurance and deterrence (Crawford, 1992).

As this emphasises, access to and behaviour in the spaces of the mall is closely controlled and monitored. Consequently, 'people walk here ... as if their conduct might be called into question at any moment' (Chaney, 1990, 64). These are effectively spaces for the affluent (white) middle classes: 'they reclaim, for the middle-class imagination, "The Street" – an idealized social space free, by virtue of private property, planning and strict control, from the inconvenience of the weather and the danger and pollution of the automobile, but most importantly from the terror of crime associated with today's urban environment' (Goss, 1993, 24).[2] While one must always look as if one has bought something or is about to buy (Shields, 1989) there are, nevertheless, those who are admitted as legitimate-looking purchasers who seek to subvert the commercial logic of the mall as a space of sale and treat it as a space to be consumed in a late twentieth-century version of *flâneurie*.

8.4   Innovatory spaces of sale in the formal economy: the mall and beyond

More generally, there has been further evolution in the built forms of urban spaces of sale, which combines three distinctive retailing forms. The speciality centre is an 'anchorless' collection of upmarket shops pursuing a specific retail and architectural theme. The downtown mega-structure, in contrast, is a self-contained complex, including retail functions, hotels, offices, restaurants, entertainment, health centres and luxury apartments: downtown malls are no longer primarily 'machines for shopping'. Now passage through the mall is an interactive experience, an adventure in winding alleys resembling the Arabian Souk or medieval town. Finally, the festival market combines shopping and entertainment with an idealised version of historical urban community and the street market, typically in a restored waterfront district (Goss, 1992). While this trichotomy is based on developments in North America, similar tendencies are visible in other parts of the late modern world. For example, there are several examples of speciality centres and festival market places in the UK. These new urban landscapes combine elements of spaces of sale with the consumption of space and space-based consumption of varied sorts. Often their significance seems to be at least as much in terms of 'urban boosterism' and inter-urban competition for investment (Harvey, 1989) or as key components of urban and regional regeneration packages (Hudson, 1994b) than as spaces of sale *per se*.

Other new innovatory spaces of sale also blur the boundaries between retail sale and other activities (Fernie, 1998). First, warehouse clubs sell food and non-food items to club members at low prices in basic 'no frills' surroundings, and factory outlet centres sell cut-price merchandise direct from manufacturers either in retail units annexed to factories or in purpose-built centres. The former have been more prominent in the USA, the latter in the UK, where they have often been linked to urban and regional regeneration schemes. They have often become tourist attractions, as this can expedite the process of gaining planning permission, as well as attracting state financial inducements (for example, at Jackson's Landing at Hartlepool Quay in north-east England). Secondly, there has been a marked development of spaces of sale in airport terminals, originally linked to duty and tax exemptions on particular categories of goods (perfumes, tobacco, wines and spirits), although now encompassing a much wider range of commodities targeted at passengers. These, by definition, are spaces of intense surveillance and, for the most part, very secure spaces of sale. Thirdly, there has been a proliferation of other 'hybrid' spaces of sale, such as railway stations and petrol/service stations, selling a range of commodities, mainly but by no means exclusively targeted at travellers. Even hospitals have become spaces of sale. These developments are indicative of the way in which the boundaries between closed, dedicated spaces of sale and the more open and public spaces of the street are becoming blurred, since they are not amenable to the same degree of control and surveillance as malls, let alone airports.

8.5 The street: a hybrid space

Long before the invention of the department store and the mall, streets formed spaces of exchange in the mainstream formal economy, 'structured and skilful spaces, a kind of classroom … in which people learned about commodities, styles and their uses and meanings' (Glennie and Thrift, 1996b, 227). Streets continue to perform as spaces of sale in the contemporary economy. Furthermore, particulars streets can become central to the emergence of new and specific spaces of exchange. For example, parts of Soho in London became consciously designated (Queer Street) as a space for homosexuals, 'a testament to the growing commercialisation of homosexuality' (Mort, 1996, 164). Evidently, resistance to dominant cultural norms may form fertile ground in which mainstream practices of exchange and the logic of capitalist social relations can flourish via creating new spaces of sale. In short, streets are complex public spaces, characterised by a multiplicity of uses, of which retail sale is only one. Indeed, the juxtaposition of a variety of people on the street engendered the perceived fear of crime and threat to personal safety, especially to middle-class women, that stimulated the creation of department stores, malls and other centres as controlled and safe spaces of sale.

There is, then, a great variety of types of street. Fyfe (1998, 2) refers to a journey 'down the broad boulevards and high speed expressways, through communities where residents participate in "daily street ballets" and on to "mean streets" where an underclass fight for survival'. However, much of the discussion of streets draws upon an extremely narrow range of high-profile studies (by Walther Benjamin, Le Corbusier, Jane Jacobs and Mike Davis, for example) of very particular and atypical streets, situated within an equally limited range of cultural and spatio-temporal settings. As a result, it reveals relatively little about the variety of streets, or the mundane and ordinary everyday street, and the socio-spatial significance of such streets as spaces of sale.

The agglomeration of particular types of shop and store in particular streets and urban quarters is a well-established aspect of major cities within the capitalist world.[3] Shops and stores selling similar commodities cluster together to benefit from agglomeration economies. However, they also cluster together in specific streets for other reasons, linked more to fashion, reputation and style, and which mix particular types of shop with a variety of other service establishments (restaurants, cafés and so on). These become both spaces of sale and spaces to be seen in and to be consumed, high-profile 'branded' streets (Fyfe, 1998). As such, the creation of such spaces is effectively a strategy of market segmentation, aimed at affluent consumers who can afford the premium prices that come with internationally branded goods, stores and streets. Major international fashion designers implement clear locational strategies in central London to create certain streets as specialist up-market spaces of sale (Crewe, 2000, 277) that appeal to the cultural, emotional and symbolic connotations of exchange and not simply the narrow economics of buying and selling. Led by 'pioneering' retailers, this involves the social construction of differentiated spaces of sale through quite precise locational preferences, based around questions of image and identity (Crewe and Lowe, 1995). For example, 43 of 50 foreign fashion designer outlets (such as Armani, Calvin Klein, Ralph Lauren and Tommy Hilfiger) in central London are concentrated in and around Bond Street in Mayfair and Sloane Street in Knightsbridge. Such streets are found elsewhere: for example, Oxford's Little Clarendon Street and Cheltenham's Promenade 'appear to have specific cultures and images attached to them – they are effectively "branded streets" and this branding enables them to attract and maintain upmarket retailers' (Wrigley and Lowe, 2002, 196). In this way, retail spaces become endowed with particular identities, which in turn help shape the identities of those who are seen there. Such addresses carry a cachet, a brand, not enjoyed by their neighbours (Fernie et al., 1997). The positive connection of product image to specific spaces in turn confers a competitive advantage and allows monopoly premium rents to be charged as issues of culture and economy co-determine one another.

Equally, retailers seek to (re)create specialised spaces of sale in particular streets via partial and selective reappropriation and discursive and symbolic reconstructions of them. Particular shops, streets and quarters are recognised as important spaces of sale precisely because they have become associated with longing for the past, memory, and nostalgia within popular imaginations. The resurgence of Carnaby Street, and the Kings Road in London, iconic symbols of the 'swinging sixties' in the UK, exemplifies the point. However, this is not simply a re-creation or repetition of the past. The re-imagination and reappropriation of Carnaby Street and the Kings Road also reflect media and UK government discourses around Cool Britannia, Britpop and the re-emergence of London as the chic global city (Crewe, 2000, 277). This renewed significance of the particular characteristics of such streets and urban quarters is expressive of resistance to, and reaction against, a tendency towards a homogenising, serially repetitive global fashion culture. There is a revived emphasis on local uniqueness as shopping streets seek to differentiate themselves from one another, playing up particular cultural and historical associations.

8.6   Markets, fairs and informal spaces of sale

While the street can become a critical space of sale in the mainstream economy, it can also form a space of resistance to the dominant logic of retailing and exchange. This is especially so of the 'ordinary' street, which can become a space filled by the informal trading and exchange practices of street traders and peripatetic market vendors. There is a long history of periodic markets in pre-capitalist exchange societies, which continue in contemporary capitalism in varied forms. They form part of a wider trend to informal spaces of sale, on the margins of the mainstream, often tied to 'second hand' goods and recycling – for example, jumble sales, flea markets, charity shops and retro-vintage clothes shops. These sites can also be seen as part of a broader move to explore alternative forms of economic relations: for example, exchange regulated by barter rather than money, that seem to hark back to 'pre-modern' times as well as to modern concerns with more sustainable economic practices. Before the establishment of shops and stores as discrete, fixed spaces of sale, many purchasers bought goods in open markets and fairs, or directly from artisan producers, or from hawkers and chapmen. Consumers of all social strata routinely acquired goods in face-to-face interaction with vendors in public settings (Glennie and Thrift, 1996b, 228). Furthermore, as shops became more established and common, many shopkeepers held stores in weekly markets and itinerant casual trading was an important mode of sale within the space of the street. Such activities continue in the contemporary era, especially via informal activity in major cities all around the late modern world. For example, in New York

immigrant street peddlers recreate a bit of the experience of Third World street markets and stalls. They also engage in less sanctioned informal markets. They join a street economy in legal and illegal goods already flourishing in poor areas of the city. Some sell stolen or pilfered goods. Poor Russian immigrants stand around ... with shopping bags full of their possessions hoping to barter or sell'. (Zukin, 1995, 210–11)[4]

FIGURE 8.2 Street markets: Sarlat, France

More recently, new forms of alternative spaces of exchange and sale, such as car boot sales in the United Kingdom, have emerged, periodic spaces of sale in which second-hand goods are (re)sold and recycled to new users, having exhausted their use value for their original purchasers. Car boot sales constitute a space of exchange in which the accepted conventions of exchange are 'quite literally turned upside down'. The disciplined social order of fixed prices, the non-contestable, non-negotiable social relations of retailer–salesperson–consumer and the trading regulations designed to protect 'the consumer' are all suspended as the consumer transforms into a hybrid entity – vendor, buyer, stroller, gazer, even entertainer. This is a space of exchange constructed by 'consumers', supposedly the subordinates to retail capital and its representatives, a space which captures the notion of 'festive life' (Gregson and Crewe, 1997).[5] As such, the car boot sale forms a 'consumption site … which synthesises leisure with consumption in a new spatial form. A place for buying and selling and a site for pleasure, somewhere to look, to wander, to rummage' (Gregson and Crewe, 1994, 266). In car boot sales much of the pleasure gained from participation stems from their intrinsically social character, for both buyers and sellers (Crewe and Gregson, 1998). The importance of sociality in exchange can be traced back to the historic role of markets as sites of communication and social exchange. Vendors find the freedom, flexibility and easy-going entrepreneurialism of the car boot sale attractive. Barriers to entry are low, social exchanges are important and work becomes indistinguishable from leisure. For buyers, participating in car boot sales is a commercial activity in which socially tactile interaction and tribal solidarity as part of the crowd is centrally important. As in traditional markets, exploring the alternative landscape of the car boot sale reflects not only a desire to acquire products but also the imaginative, sensory experiences that it promises.

Consideration of spaces of second-hand and informal exchange and the tracking of products though various cycles of use and re-use provides a useful corrective to accounts that prioritise single acts of exchange in formal spaces of sale. Such spaces and practices also reveal the importance of commodity recycling (in the process raising questions about what we mean by use value), ethical consumption practices and consumption motivations centred on notions of thrift and the bargain. There are very large numbers of poor people involved in a huge, informal trade in second-hand clothing and household items through such spaces through economic necessity rather than ethical choice, for example. Current moves towards more ethical and sustainable forms of production and consumption are increasingly meaning that 'the regime of consuming subjectivities is to be the target of a critique, its contradictions exposed, the hidden costs – individual, political and cultural – of its surface pleasures revealed' (Miller and Rose, 1997, 2).

8.7   The home

The home is a multifunctional space that, *inter alia,* can function as a space of
sale. Shopping from home originated in the USA in the late nineteenth century,
as mail order from catalogues allowed the consumer revolution to be diffused
to isolated rural populations. The catalogue provided rural people with
'a department store in a book' (Schlereth, 1989). By the mid-twentieth century
in the UK, mail-order firms such as Freeman, Grattan and Littlewood had
emerged as major mail-order retailers. Typically, their goods were sold via
women, who acted as sales agents to working-class people living in their
neighbourhood. Customers were attracted by the combination of spreading
(weekly) payments over a number of months and the delivery of goods to the
home. Women sales agents were attracted by this work as it enabled them to
work from home (payment was via a commission on sales) and so was compatible with a particular gender division of labour that required their caring
for children or relatives or their meeting the domestic needs of husbands and
sons engaged in shift work in mines and factories. By 1970 over 4% of all retail
sales were by mail order, with much higher shares in product segments such as
clothing for women and children and soft furnishings (McGoldrick, 1991). The
aggregate share fell to around 3% by the late 1980s/early 1990s and also fell
sharply in the USA, symbolised by Sears Roebuck, one of the original mail-
order firms, closing its mail-order catalogue business in 1993.

As the market share of the established mail-order firms fell, they shifted
their approach, leading to expansion in new types of catalogue-based sale from
the home. This involved a shift from agency catalogues to direct marketing
catalogues and 'specialogues', directed at specific lifestyle groups and market
segments of potential purchasers. Major direct marketing companies, such as
Lands' End, Next and Racing Green, relentlessly target specific market niches
and social groups, especially dual income households and busy professionals,
identified via accessing electronic databases on their incomes and past purchasing patterns. Established catalogue sales companies, such as Sears (which
re-entered the sector via joint ventures), Spiegel and Talbot, have also adopted
their operations to produce a more focused and targeted approach (Wrigley and
Lowe, 2002, 236–7). This has been instrumental in the recent expansion of 'non-
store retailing' (Christopherson, 1996), largely involving payment by credit card.
This expansion in more up-market niche catalogues is linked to growth in dis-
posable incomes and the proliferation of lifestyle projects but the growth of
other catalogue sales reflects different motives. For example, in the UK both
local (for example, Loot) and national (for example, Argos) catalogues reveal a
shared appeal to social groups precluded from expensive high street shopping.
Successful use of such distanciated spaces of sale is predicated upon the development and acquisition of appropriate skills and knowledges. For 'finding a bargain among an array of unseen goods whose product specifications are described
by the vendor … requires considerable skill, risk and time' (Clarke, 1997a,

78–9). However, the mis-match in skill and knowledge of purchaser and vendor create the ambiguity of an unregulated market place and the potential for disappointment as well as for bargains. A perennial problem with all catalogue-based buying from home is that purchase decisions are based at best on small photographs and limited written descriptions of the product. Consequently, potential purchasers are often dissatisfied when they actually receive the product and abandon the purchase.

More recently, shopping from home via television and the Internet has expanded in some parts of the contemporary world. This form of shopping requires considerable economic and technological resources, both within the home and in terms of networks connecting the home with other locations. Television shopping is most developed in the USA, with dedicated TV shopping channels on which various commodities are demonstrated, leading to telephone orders. By 1998 some 10% of consumers in the USA were purchasing in this way. These were mainly lower income, older female customers, with most purchases concentrated on cheap jewellery, women's clothing and personal care products (and generally eschewing nationally recognised brands). QVC, the leading operator, established a home shopping network in the UK (Wrigley and Lowe, 2002, 238–9) but it has had a limited impact. The introduction of more interactive TV systems, allowing potential purchasers to browse and request advice and information, may enhance use of this medium for shopping from home. However, future growth will more likely be based on the Internet and cyberspace as a space of sale that can be accessed from home and this is discussed more fully in the next section. Currently, then, shopping from home can be based upon a variety of media that connect the potential purchaser with vendors – window shopping from catalogues, television and the Internet, for example (Crewe, 2000, 278–9) – expanding the potential of the home as a space of sale. In the mid-1990s 60% of all direct sales in the USA took place in the home (Berman and Evans, 1998, cited in Wrigley and Lowe, 2002, 243).

There is, however, an important distinction to be drawn between shopping *from* home and shopping *at* home. Shopping at home involves direct face-to-face contact between vendors and purchasers in their home. More often than not, agents acting on behalf of companies (a predominantly part-time retail workforce, not unlike catalogue agents) perform the work of selling commodities such as cosmetics and perfumes, cleaning products, cooking and kitchenware, encyclopaedias and jewellery. While such direct selling may involve a degree of sociability, another variant of direct sales in the home is predicated *upon* sociability. The 'party plan' system is most famously linked with Tupperware, first in the USA and then in the UK, in the 1950s and 1960s (although subsequently deployed by other companies such as The Body Shop and Ann Summers). It is based on the proposition that salespeople encourage customers to act as hosts. They invite friends and neighbours to a 'party' to play games and consume refreshments while products are displayed and demonstrated (Clarke, 1997b).

As well as serving as a highly rarefied sales forum, the party acts as a ritual interface between maker, buyer and user. The 'hostess' offers the intimacy of her home and the range of her social relations with other women (relatives, friends, neighbours) to the Tupperware dealer in exchange for a gift. The dealer, supplied by an area distributor, uses the space to display products and gain commission accrued on sales. In addition, she can arrange further parties from among the hostess's guests. For the purposes of 'the party', the home becomes a deeply gendered, and in some ways oppositional, space of sale, since it sanctions all-female gatherings outside the family. Loyalty to fellow neighbours and friends is the linchpin for attendance to many parties and for many women it was and is an opportunity to socialise outside the home at little expense. While the pretext of the gatherings is domestic, this does not preclude women from directing the conversation and interaction towards other concerns. While substantiating predominantly conservative and traditional feminine roles, the Tupperware party also provides a pragmatic pro-active alternative to domestic subordination.[6]

In bringing the business of selling so prominently into the space of the home, the concept of 'the party' emphasises the way in which any form of selling based in or from the home blurs the boundaries between the private space of domesticity and retreat from the social relations of capital and the market as a public space of sale (or as Clarke puts it, the intrusion of the 'market' into the sanctity of 'domesticity'). Certainly there is a class dimension to the more recent ways in which this blurring comes about since it is limited to the homes of those who can afford the enabling technologies. Whatever the mode of interpenetration, however, the living room and kitchen become spaces of sale, located deep in the heart of the home. While this can be seen as disturbing domestic bliss and the sanctity of the home, for many women the home was a space of work and confinement rather than some idyll to be disturbed by the penetration of market relations. For them, catalogue selling or 'parties' may provide welcome sociability.

8.8 Cyberspace and electronic spaces of sale: rhetoric and reality

Sherman (1985), reflecting on the growth of information technology and home computing, presciently observed that the home appeared destined to take on new functions as an entertainment, shopping and banking centre. The growth of the Internet and B2C commerce based on the purchasers' home has further extended developments in electronic commerce. By the end of the 1980s major retailers were using EPOS (electronic point of sale) data to automate and control linkages between within-store inventory, the warehouse and distribution network, and central administrative functions such as accounting, analysis, ordering and purchasing. These systems were increasingly connected via EDI (electronic data exchange) into the computer systems of manufacturers and suppliers, allowing 'paperless' supply chain control via electronic exchange of invoices, orders and so on (leading to massive reductions in the employment of clerical labour). EDI also facilitated electronic tracking systems (retailer interrogation of manufacturers', suppliers', contractors' or distributors' computer systems) to discover the whereabouts of a particular order in the supply chain. This permitted further shortening of order cycle times and significantly increased predictability and delivery lead times. For example, by the late 1990s within food retailing, major stores could be supplied smoothly and predictably within 12 hours of an order being automatically transmitted, compared to 48 hours in the late 1970s (Lowe and Wrigley, 1996, 10), while allowing supply chains to be stretched further over space (Crewe and Lowe, 1996, 274). The development of these electronic spaces and systems of ordering also permitted, and required, changes in the spatial configuration of retailing centres, with new large out-of-town stores that allowed easy access for regular delivery.

In addition to these sophisticated intra- and inter-company electronic database supply chain management systems, many consumers had become accustomed to making telephone transactions (as the costs of long distance telephone calls fell). Processing of many routine transactions had been centralised in call centres, especially for the purchase of products such as tickets and computer-related items and for services such as banking and bill payments. The provision of such services became 'dis-intermediated', replacing workers in call centres with digital links. The growing number of households owning PCs, increased familiarity with e-mail, the development of inexpensive user-friendly browsers and increasing connections to high speed, broad band links, created a potential pool of B2C purchasers. The extension of such activity into interactive cyberspace 'simply' required their habituation to purchase there (Kenney and Curry, 2001, 59). However, this remains a very socially and spatially restricted pool because of the uneven diffusion of home PCs.

B2C systems of electronic commerce shift the point of sale to the purchaser's home or other points of access to the Internet such as cyber cafés or local corner stores (as in Japan: Aoyama, 2001). This is a critical necessary precondition for purchasing in this way, for e-commerce and electronic spaces of sale require a material basis in terms of hardware (PCs, broadband connections) and software (code, programs and websites, for example). For those who can purchase in this way, however, there can be considerable gains in convenience and savings in time and travel costs.[7] There has also been significant growth in Internet auction markets (C2C commerce), especially of eBay, with 350,000 lots offered for sale every 24 hours (Leinbach, 2001, 23).

These new electronic channels allow vendors precisely to target specific market segments and niches, although rhetoric has run far ahead of reality in terms of the predicted growth of cybersales and e-commerce. The collapse of many dot.coms in 2000/2001 is one reason for the 'mounting evidence in many sectors of market concentration', with 75% of B2C and C2C e-commerce in the USA transacted through five sites – Amazon, eBay, AOL, Yahoo!, Buy.com (Button and Taylor, 2001, 30). Moreover, the growth potential of web-based sales may be limited to a few specific commodity categories, such as books and consumer electronics. Books are ideal commodities for this type of sale as, unusually, it makes no difference whether you buy your copy of *Das Kapital* from Amazon.com or your local corner book shop – it will be the same book, irrespective of where and how it is purchased. Even so, by 2000 online sales constituted less than 1% of total book sales worldwide. Thus while there may be 'remarkable benefits' for retailers in transferring sales to the Internet (Kenney and Curry, 2001, 58), for example in lowering inventories and reducing fixed capital investment costs in shops, these vary by product and service and such benefits are limited to a very few cases. Assertions that 'the sale of physical goods via the Internet has become incredibly important for all retailers' (Leinbach, 2001, 22) are simply incorrect.

Partly for these economic reasons, partly because e-commerce denies the opportunity for sociality in buying and selling, partly because there are similar risks of disappointment associated with catalogue sales and TV shopping (Kitchen, 1998), there is an increasing tendency for the development of hybrid spaces of sale ('bricks and clicks'). These combine the benefits of cyberspace ('clicks') and more conventional material spaces of sale, such as shops, stores and malls ('bricks'). Furthermore, the tendency to create such hybrid spaces reflects the growing interest of major retailers in e-commerce to supplement the sales of their conventional shops and stores, rather than an evolution from the original dot.coms. For example, in 2000 McDonald's acquired Food.com, an Internet food takeout and delivery service, aiming to become the preferred web destination 'for anything to do with food' (Westwood, 2001). The satisfactory purchase of many commodities (for example, clothing and household furnishings) requires seeing, feeling, touching and smelling them to ensure that they are aesthetically pleasing. In other cases, such as shopping in discount stores, leisure shopping becomes a quasi-sport, namely, hunting for bargains.

However, while online shopping eliminates the experiential aspect of shopping, it may widen potential choices. Potential purchasers can scan websites from the comfort of their own home, identify commodities of interest to them, and then visit specific shops to inspect and, possibly, purchase them. Thus the Internet becomes a new search technology rather than an electronic space of sale *per se*. People 'frequently use the Internet to gather information about products, and depending on the perceived reliability of product quality and suitability, many are still purchasing off-line' (Button and Taylor, 2001, 35). In doing so, they tend to rely on a small number of portals, one-stop shops that are central nodes in cyberspace and seek to meet all the needs of Internet shoppers. Habitual use decreases the probability of moving to another portal because of the costs of moving, re-entering data and becoming familiar with a new interface (Kenney and Curry, 2001, 61).[8]

While strong claims have been made about e-commerce radically reconfiguring spaces of sale, there are limits to the capacity of any new technology to overcome the contradictions of the social relations of capital. Such technology is produced within those social relationships and the capability to use it is similarly circumscribed. The capacity to join the Internet depends upon the ability to purchase hardware and software produced as commodities, and the impacts of this on uneven access are readily apparent. Globally, very few homes have the required Internet connections, with the majority of these in the USA (although there are rapid growth rates in other parts of the world), but it is possible to access such sites via computers in more public spaces such as Internet cafés and public libraries. Even so, Internet access is very uneven, with sharp class and income divides. Moreover, there is no guarantee that greater access or even a dramatic fall in the real costs of hardware and software would lead to substantial increases in e-commerce at the expense of more conventional forms and spaces of sale. There are numerous other barriers to e-commerce expansion. From the point of view of retailers, e-commerce must meet the requirements of profitable sale. The evidence to date suggests that, with specific exceptions, the impact of B2C and C2C e-commerce is to reinforce or modify existing patterns of purchase and spaces of sale rather than radically alter them.

8.9 Spaces of sale, spaces of work

Spaces of sale are spaces of work performed by both buyers and sellers. There is a variety of types of 'selling work', depending on whether a space of sale is within the formal, informal or illegal economies, whether sale occurs in physical space or cyberspace, and upon the character of the commodities – goods and services – being exchanged. The diversity of work involved in selling requires a variety of approaches to maintaining the 'frontier of control' in the workplace and regulating the labour process. Moreover, trends in work can be contradictory, with simultaneous emphases towards deskilling and reskilling within spaces of sale.

Retailing is a 'leading edge' industry in terms of transforming labour practices in the advanced capitalist world (Lowe and Wrigley, 1996, 11). Despite high levels of capital investment, labour costs remain very significant. The need to be near to customers severely circumscribes the extent to which companies can re-locate in search of cheaper labour. Nevertheless, 'many of the tasks involved in selling commodities have been re-designed and combined so as to decrease labour inputs or direct them to serving the most profitable market segments' (Christopherson, 1996, 159). One response has been an increasing reliance on self-service, especially in the sale of more routine and mundane commodities. Self-service retailing reduces the specialisation previously required in shop work, downgrading the majority of tasks within larger stores to shelf-filling or till operation. Another response has been to substitute contingent, part-time and temporary for full-time and permanent workers. Retailing has been the site of important innovations in managerial control of the labour process. For example, via the skilled use of a multiply segmented contingent labour force and the increasingly sophisticated use of EPOS-based IT systems to monitor workers and deploy workers, work hours and work practices very precisely in relation to temporal fluctuations in demand. Following labour market deregulation in the 1980s, for instance, multiple retailing chains in the United Kingdom, especially in food, increasingly employed part-time workers to match employee levels to temporal (hourly, daily, weekly, seasonal) fluctuations in consumer demand.

While some spaces of sale have therefore been characterised by a long-term tendency towards a deskilling of work, other new sorts of employment are emerging – or at least becoming more prevalent – elsewhere. For example, there has been a proliferation of new jobs in call centres scattered around the globe. Within conventional shops, some of the new jobs are seen as charac-terised by a greater degree of worker autonomy, empowerment and responsi-bility ceded to skilled and knowledgeable workers, who are required to enact and perform the work of selling in particular ways, and often to assume a par-ticular bodily appearance. An embodied performance is increasingly signifi-cant in a range of selling occupations (McDowell, 2001, 241). As such, sales assistants increasingly comprise the actual product on sale. This 'selling of the staff' is most pronounced in 'customer care' strategies, which prioritise customer service as the key determinant in contemporary strategies of retail competition (Lowe and Wrigley, 1996, 24). Furthermore, purchasers often wish to be served by people who are 'like them'. Consequently, certain retail employers consider youth or ethnicity to be important attributes over and above aptitude or retail experience, with black workers being particularly concentrated in high-fashion stores with a strong image-consciousness. Sometimes, however, it appears that what is involved is a reinterpretation of the characteristics of well-established jobs rather than the emergence of new and qualitatively different forms of employment. For example McRobbie (1997, 87) argues that aspects of performance and presentation involved in the work of selling in fashion retailing are both specific to that activity, integral to the job and tied to the appearance of the person performing the work. However, this is hardly a new requirement of sellers of fashion. Furthermore, in so far as such work necessarily depends upon the appearance and personal attributes of the worker, it is linked to age and stage in life cycle and so limited in its duration for an individual. As such, fashion retail workers' 'self-image must surely be undercut by the reality of knowing that in a few years time, possibly with children to support, it is unlikely that they would hold onto the job of deco-rating the shopfloor at Donna Karan' (McRobbie, 1997, 87).

8.10 Summary and Conclusions

In this Chapter I have sketched out the variety of socio-spatial forms of spaces of sale, in the formal economy, the informal economy and in the home. There has been, and is, considerable spatio-temporal variability in these forms and in the ways in which they have been constructed, materially, socially and discursively. The configuration and manipulation of retail space is an intensely geographical phenomenon. Do such spaces of sale reflect consumer sovereignty and the revealed preferences of consumers? Or do they seek to construct consumers as dupes, hapless victims conned into purchases by an environment designed to increase (in the terminology of mall architects) 'dwell time'? Have developers and designers of the retail built environment consistently exploited an intuitive understanding of the structuration of space in order to mould the retail environment so as to sell (more) commodities? Or, in contrast, are those who populate these spaces of intended sale knowledgeable people who understand the purposes of the mall designers but actively seek to contest and subvert them, seeking to use them in unauthorised ways? For example, they perform as 'post-shoppers', hanging out in the mall with no intention to buy.

While spaces of sale vary in form, there are tendencies towards the serially monotonous production of identikit shops, streets and malls 'in which you could be almost anywhere' (Goss, 1993, 32). This tendency is linked to attempts to homogenise (global) consumer preferences, tastes and purchasing patterns through the internationalisation strategies of global retailers. At the same time, however, there are counter-tendencies that re-emphasise the importance of specificity and variability in spaces of sale and the exclusivity, uniqueness and quality of the commodities and goods on offer in them. Often this spatial branding of commodities is associated with the cultural and historical attributes of such spaces, seeking to create a space to consume as well as a space of sale. Beyond such varied spaces of sale in the mainstream capitalist economy, the home, street markets and car boot sales constitute spaces that are partly in the mainstream formal economy and partly in the informal economy. Their presence further emphasises the variety of spaces in which exchange and sale occur. While selling is a service and many spaces of sale require the time/space co-presence of buyer and seller, whether in physical space or cyberspace, this is not *necessary*, as examples such as catalogue sales make clear.

Notes

1 There are also spaces of wholesaling, requiring specific configurations of fixed capital. However, the emphasis here is on spaces of retail sale and sales to final consumers.

2 The proliferation of CCTV cameras in city centres in the UK signals an attempt to make these public spaces more like the private spaces of the malls.

3 See the models of urban retail structure developed by geographers (Berry, 1967).

4 Such activities are not confined to major cities. During the 1990s in the northern Greek town of Kavalla, there were regular weekly markets at which Russian immigrants sought to sell their possessions to local residents or passing tourists.

5   Whether such spaces are 'constructed by consumers' is, however, debatable, but the social relations of their construction are clearly very different from those involved in the material and discursive construction of major malls and department stores.

6   In 2003, Tupperware parties in the UK ended. Invitations no longer had their former 'social cachet' (Clarke, cited in Hamilton, 2003), while lifestyles had changed. However, they remain popular in the USA and have grown in popularity in emerging capitalist markets in the 1990s so that, globally, a new Tupperware party begins every 2.6 seconds.

7   By 1999 30% of homes in the USA had Internet access. More than 17% had used the Internet to purchase goods and services (Button and Taylor, 2001, 28).

8   The success of such portals underlies the failure of e-malls, the electronic analogue of the physical shopping mall, since these focused simply on retail sales rather than provision of a wider range of services.

9 Spaces of Consumption, Meaning and Identities

9.1 Introduction

Consumption is 'an ongoing process rather than a momentary act of purchase' (Crewe, 2001, 280), those activities that follow purchase. The fixing of capital in particular ensembles and spatial forms – including houses, schools, hospitals, shopping centres – creates the material spaces of consumption. This produces a range of public, private but open to the public, and private (home) spheres and spaces in and/or through which consumption is performed. For example, golf courses and roads constitute spaces for the consumption of golf clubs and automobiles, respectively, while consuming spaces is an integral experiential part of the holiday, along with the consumption of food, drink, automobiles, golf clubs and locally produced services in those spaces. Spaces of consumption are also spaces of work in which specific forms of service work are carried out and performances enacted. They may be private (for example, a caterer or hairdresser comes to your home), but more usually are 'public'/private (for example, you go to the hairdresser or the restaurant). Furthermore, consumption is also linked to the creation of identities. This is particularly so given the increasing commodification of the lifeworld and the aestheticisation of daily life, the growing importance of the symbolic and sign aspects of the commodity form, and the socially ascribed meanings of brands.

9.2 Spaces of consumption, display and identity formation

9.2.1 The street

While streets perform as spaces of sale, they simultaneously constitute spaces consumed by a variety of people, many not engaged in the business of buying and selling. Streets form sites of display, spaces of leisure and recreation in which to stroll and see, exemplified in the activities of the nineteenth-century middle-class *flâneur*. This continues in the contemporary era, although the character, identities and motivations of *flâneurs* have changed. For Mort (1996, 176) the 1980s *flâneur* of London is a homosexual man 'cruising the streets with a clear agenda'. He (1996, 164) links this to the ways in which parts of Soho became consciously designated as a space for homosexuals (Queer Street) via 'the deliberate attempt to fuse together a new upsurge of radical sexual politics with the celebratory style of the street festival'. This connects issues of identity and (self-)image to the ways in which different people define and consume the public space of the street and in which the street, as public space, can become a site of resistance to dominant cultural norms. In this sense, Soho can be seen as the latest in a venerable line of cultural movements centred on space-specific resistance to such norms. That it did so while becoming a new space of sale within the mainstream economy, exploiting emergent issues of cultural and sexual identity, exemplifies the contradictory character of this resistance.

While the activities of *flâneurs* reveal that the street is a space consumed by the middle classes, streets are not a middle-class preserve (not least as *flâneurie* has ceased to be a middle-class, middle-aged and male pastime). The nineteenth-century urban street 'belonged in large part to labouring-class people. For poor men, women and children, the streets were workplaces and playgrounds' (Glennie and Thrift, 1996b, 227). This remains the case today, with the street an important public space. Nevertheless, everywhere there are sanitised middle- and upper-class residential streets, cleansed of categories of people designated as socially undesirable, denying the street as a public space of consumption to those categorised as 'undesirables'.

9.2.2   *The department store*

Department stores rapidly became key elements of the modern urban environment from the late nineteenth century, focal points for social life, especially for middle-class women. Indeed, the rise of the department store can be seen as a process of feminisation of the *flâneur*, by creating a 'public space' insulated from the dangers of the street and safe for women (Featherstone, 1998). A central theme of the department store as a new 'modern' retailing space was its role as a centre of entertainment and as a tourist attraction. It formed a space to be consumed, combining shopping with the mobilities of tourism (Goss, 1999).[1] For example, Selfridge's soon became 'one of the great show sights of London', designed to become a 'social centre' in which people could browse and meet; the first advertising campaign urged customers to 'spend the day at Selfridge's', emphasising that there was 'no obligation to buy' (Nava, 1997, 69).

Similar but intensified processes are observable within contemporary department stores, which 'combine within their "spaces" designer interiors, rituals of display and leisure, sexuality and food' (Lowe and Wrigley, 1996, 25). With developments in advertising techniques, and a growing emphasis upon the aesthetic and symbolic attributes of commodities as many goods became increasingly standardised, the significance of the activity of shopping and the actual space of purchase of the department store in constructing consumers' identities has been enhanced. Purchasing there, in *that* store, became correspondingly more important in conferring social status on the purchaser and affirming her/his identity (Chaney, 1983). Nonetheless, the prime rationale of such stores is as spaces of sale, not as spaces of identity formation *per se*.

9.2.3 The mall

Prior to the mid-1950s in the USA malls sought to attract purchasers on the basis of easy access by automobile and free parking. Subsequently, mall managers moved away from this modernist emphasis on convenience and 'rational' shopping towards the promotion of these spaces on the basis of their 'carnivalesque atmosphere', a carefully controlled and sanitised quasi-street (Lash and Urry, 1994, 235). Furthermore, mall developers 'rediscovered the appeal of the turn of the century department stores, transforming indoor spaces into theatrical "sets" in which a form of retail drama could occur' (Hannigan, 1998, 90). This rediscovery was first materialised in Southdale Mall, Minneapolis, with the 'garden court of Perpetual Spring', an atrium filled with exotic flowers and plants that bloomed all year round, as its focal point. From this time, the emphasis increasingly switched to the mall as a space to be consumed and not simply a space of sale.

This tendency towards 'de-differentiation, designing the mall as a space of entertainment, leisure and a tourist attraction to be consumed as much as a space for the instrumental purchase of commodities, became intensified with the 1980s development of the West Edmonton Mall. The WEM typifies processes of 'disneyfication' and 'imagineering', 'a spectacle integrated into the everyday and open all year round'. Indeed, 'the annexation of much of a city's retail/social life into a corporate, self-contained "disneyfied" built environment ... is unparalleled'. However, to describe the WEM as a 'disneyfied' and 'imagineered' shopping mall 'is a static analogy that is metaphorically correct but reveals nothing about process(es). ... WEM is neither merely disneyfied nor Disneyland simply imagineered, both are part of a much larger set of processes, one of which is simulated everywhereness'. This involves the 'overt manipulation of time and/or space to simulate or evoke experiences of other places' (Hopkins, 1990, 2).

Although translated to spaces beyond the USA (such as the Bluewater Centre in Kent and the Gateshead MetroCentre, both in England), this tendency reached its (as yet) apotheosis in the USA with the development of the Mall of America, in Bloomington, Minnesota. The Mall is the most visited tourist attraction in the USA, a sophisticated themed environment that speaks to a romanticised version of street life in the city. It is divided into districts, based on tourist retail destinations, drawn from different parts of the world, while

concept restaurants and fast-food outlets are positioned as imaginary destinations in the consumer-tourist worlds. Rainforest Café, for example, is described as an 'enchanted place for fun far away, that's just beyond your doorstep'. Diners are issued 'passports' for a meal that is a 'safari' and are exhorted to complete their 'trip' by buying a souvenir stuffed-animal toy in the adjacent 'retail Village'. (Goss, 1999, 54)

Within the 'kinaesthetic space' of the Mall,

> contemporary *flâneurs* are literally moved to aggregate, shop and celebrate: they
> are drawn across thresholds and along paths by the use of contrasts in color and
> light, focal attractions and linear design elements; carried away by escalators
> and elevators; directed by spatialized narratives in the form of waymarkers and
> sequential interpretative texts [in an] aesthetics of motion. (ibid., 52)

More recently, the boundaries between shopping, leisure and entertainment
have been further blurred, with the emergence of 'shopertainment', a 'hybrid'
fusion of shopping and entertainment. In 'the theme park cities of the 1990s,
shopping, fantasy and fun have further bonded in a number of ways. ... The
two activities have become part of the same loop: shopping has become intensely
entertaining and this in turn encourages more shopping. Furthermore, theme
parks have begun to function as "disguised market places"' (Hannigan, 1998, 92).
One variant of this themed retail experience, 'experiential retailing', is exemplified by Nike Town, a retail theatre showcase in New York. Opened in
November 1996, Nike's flagship store is

> a fantasy environment, one part nostalgia to two parts high-tech, and it exists to
> bedazzle the customer, to give its merchandise sex appeal and establish Nike as
> the essence not just of athletic wear but also of our culture and way of life. The
> retail element of the store is more muted: one can buy Nike products at Nike Town
> but the store exists primarily to promote brand recognition. (Hannigan, 1998, 92)

As such, Nike Town represents an important qualitative shift in the process of
selling and the fusion of shopping and entertainment for it prioritises the rein-
forcement of the brand; it is more a device for brand marketing and reinforce-
ment than one for the sale of specific commodities. Moreover, this is having a
feedback effect on older spaces of sale. 'Somewhat ironically', department stores
such as Macy's 'have begun aggressively to pursue the concept of "retail entertainment", taking their cue from speciality stores such as Old Navy, whose
"industrial chic" – concrete partitions, pumping music, bright lights and electronic messages – appealed to the important youth market' (Wrigley and Lowe,
2002, 185–6). What comes round, as they say, comes round.

For some people, however, there are strict limits to the enchantment and
magic of the mall. For them, 'the magic ... is elusive. There is a connection
between the commodities bought and the person themselves. The things they
buy carry the poignancy, the tragedy, the longing, the despair, the emotional
intensity of human relationships ... most of all they convey the relationship with
a very tragic self' (Baker, 2000, 4). They may 'de-shop' – that is, purchase goods
with the deliberate intention at the moment of purchase of returning them rather
than going on to consume. More worrying are 'those whose lives become
defined by excessive and addictive consumption and associated debt, imprisonment, and sometimes suicide' (Crewe, 2001, 635). For them, the magic of the

mall dissipates in a dystopian world of excess that defines identity in very dangerous ways and sits uneasily with the rhetoric of enchantment and magic.

However, the mall has also become a focal site of identity formation in other unintended and, to mall managers and developers, unwanted ways. Malls have become 'an essential site for communication and interaction, a place for "hanging out", for "tribalism" where adolescent sub-cultures are formed and where lifetime experiences take place' (Lowe and Wrigley, 1996, 22). The notion of 'tribe' (Maffesoli, 1992) denotes a series of loosely bonded and temporary social groups crystallised out of the mass via specific forms of sociality (Glennie and Thrift, 1996b, 226). Being in the mall and behaving in specific but non-purchasing ways becomes central to collective identity formation within such tribal sub-cultures, based upon contesting the dominant construction of the mall as a space in which identities are formed via buying and consuming. This is one reason for the strict surveillance (via CCTV, security guards and so on) of many malls and shopping centres, to try to eliminate such 'deviant' and non-conforming behaviour.

9.2.4   *Inconspicuous consumption spaces*

Crewe (2001, 278) defines inconspicuous consumption spaces as 'a range of unconventional spaces' and their associated practices, those informal spaces of sale that also function as spaces of consumption and spaces to be consumed – car boot sales, charity shops, retro-vintage clothes shops, for example. For some, purchasing second-hand reflects the economic constraints of limited purchasing power but for others there are different non-instrumental motivations, as commodities discarded by some become attractive objects to others. This transformation depends upon perceptions and knowledge of authenticity, endowments of cultural capital that form the basis for processes of discernment and distinction, and upon definitions of value and processes of valuation. These are central to those exchanges that occur as a prelude to the (renewed) consumption of commodities which, for some, are no longer deemed to have use value but for others have become desired, to be acquired and consumed (Gregson et al., 2001). Often, however, items acquired second-hand are subject to time-consuming 'divestment rituals', such as cleaning, repairing and altering, 'all with a view to expunging all traces of an unknown other'. In this way it 'becomes possible to transfer, obscure, lose or re-enchant the meanings of commodities as they pass through endless cycles of use and re-use' (Crewe, 2001, 281). Consumption becomes a social process, recursively conducted in and through a variety of circuits and spaces.

9.3   Holidays, leisure and the consumption of spaces

The consumption of spaces is integral to being a tourist, to performing tourism. Conversely, those promoting and selling tourism increasingly have sought to forge distinctive images of tourism spaces, and create differentiated tourist destinations. These spaces of tourism are constructed – materially and discursively – to appeal to particular types of tourist or visitor, who seek to consume different aspects of these spaces: culture and tradition, ecology and nature, or more prosaically, sun, sea and sand. This process of 'enchanting' commodified spaces of tourism, constructing them as different, as exotic, requires that tourists possess the requisite cultural capital to read tourism spaces in these ways. This has recursive impacts upon their knowledge and identity, so that 'many visitors are becoming increasingly skilled at evaluating landscapes and townscapes, at building up their cultural capital so as to be able to form more sophisticated aesthetic and environmental judgements'. This in turn has feedback effects upon the construction of spaces of leisure and tourism. Consequently, the increased aesthetic reflexivity of subjects in the consumption of travel and of the objects of the culture industries 'creates a vast real [that is, material – RH] economy, produces a complex network of hotels and restaurants, of art galleries, theatres, cinemas and pop concerts, of culture producers and culture "brokers", of architects and designers, of airports and airlines and so on' (Lash and Urry, 1994, 58–9).

For example, museums such as Beamish are constructed as spaces that bring together a range of artefacts originating within north-east England in which people consume regional 'history' (M. Hudson, 2001). Producing regional histories may involve the creation of fictional spaces, based upon often romanticised accounts of the past. Other museums focus on a specific industry, such as Big Pit at Blaenavon. They are examples of seeking to commodify an industrial past as part of a process of constructing an alternative economic future in and for such spaces (Hudson, 1994b). Such museums are represented as heritage sites in which 'old' things, commodities that no longer have their original use value, are re-valorised for and through their consumption by others as history, as heritage. These artefacts are materially the same but their meaning alters in the collective space of consumption of the museum, as they are presented in particular ways as part of a discourse of heritage, to be read in prescribed ways.

Rather than being driven by aesthetically aware tourists, however, the increasing differentiation of spaces of tourism and the production of particular types of tourist space may reflect the market segmentation strategies of capital and public policy makers charged with marketing such spaces. Consequently, potential tourists are bombarded with a plethora of images, a 'manufactured diversity', but have become increasingly skilled at performing semiotic work and interpreting such images. However, some consumers of tourism spaces have become less susceptible to the inducements of such images. Like the 'post-shoppers' of the mall, post-tourists contest and seek to subvert the dominant projected images of particular tourist spaces, to perform tourism to their own script rather than the scripts of travel agencies, tour operators or public sector tourist development agencies.

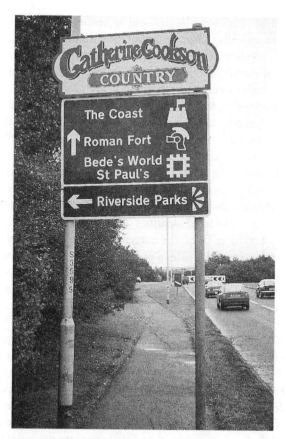

FIGURE 9.1　Creating tourism spaces from fiction: Catherine Cookson
Country, South Tyneside, England

9.4　Privatising public space: social exclusion and the contradictions between spaces of sale and spaces of consumption

Specific urban spaces are significant to the production of sociality, 'the basic everyday ways in which people relate to one another and maintain an atmosphere of normality' (Glennie and Thrift, 1996b, 225–7), constituted via relations of co-presence and co-present interactions. Strong social relations, such as those of class, ethnicity or gender, powerfully determine patterns of sociability, as, increasingly, does age. In many of the peripheries of the contemporary capitalist world, streets, market places and public gatherings are important sites of sociality. In the past, they were, and to a degree still are, also sites of sociality in the core territories of contemporary capitalism. Increasingly, however, specifically delineated spaces of retail sale, leisure and recreation, and entertainment have become significant settings for group involvement and interaction.

Processes of capitalist development and property relations have led to public spaces of sales (the street, the market) becoming at least in part replaced by privately owned, often closely monitored and carefully managed, pseudo-public spaces. They become sites for promoting festivals (like Christmas and Easter), community events, special displays and seasons and so on, carefully building up local affinities to produce 'an affective ambience' (Maffesoli, 1991, 11). They become spaces in which 'people can interact lightly without too much hanging on the outcome'. However, the crowd or throng is performative and cognitive and aesthetic, 'remodelled arenas of sociality [providing] relatively safe proving grounds for the individualising self' (Lash and Urry, 1994, 235). No longer do people go to the crowd with fixed identities (in terms of class, ethnicity, gender and place), with any deviation from them regarded as a temporary carnivalesque anomaly. Rather, the crowd provides a public space for experimentation with new identities and variants of the self, seen as routinely fluid, shifting and temporary. It is doubtful, however, whether all can participate equally, or would indeed wish to participate, in this process of mobile and fluid identity re-formation.

However, the rationale for constructing and fixing capital in malls is not experimentation in individual identities. Successful mall development depends upon sufficient money being spent there to ensure commercial success. As such, there is a tension between promoting the mall as a space of entertainment, leisure, recreation and tourism and ensuring that these activities generate revenue and help increase sales in the shops and stores within these pseudo-public spaces of sale. Consequently, managers cannot be indifferent as to who populates the space of the mall, to ensure that only legitimate potential purchasers are found there. Those who populate the mall must have both the motivation and the money to buy. Those who cannot, or will not, engage in the performance of purchase cease to be regarded as legitimate economic citizens, and as such are to be ejected to allow those who can and will to do so unhindered.

However, such attempts to privatise public space do not go uncontested. Some are admitted to the mall as legitimate-looking purchasers but nevertheless seek to subvert its commercial logic. Some engage in the activity of 'de-shopping' (see p. 170). More generally, 'the mall has its post-shoppers who, as *flâneurs*, play at being consumers in complex, self-conscious mockery'. As such, they 'actively subvert the ambitions of the mall developers by developing the insulation value of the stance of the jaded world-weary *flâneurs*; asserting their independence in a multitude of ways apart from consuming ... ' (Shields, 1989, 160). Moreover, *flâneurie* has acquired a less gendered character, becoming commonplace among new middle-class consumers. They wander, they hang out, they look and window-shop, they observe the crowds and the throng, they derive enjoyment from doing so, but credit cards, cheque books and cash remain firmly out of sight. However, because they always look as if they have bought something or are about to buy, they manage to remain. In contrast, others who do not fit the template of prima facie credible purchasers and whose motivation for visiting the mall is to socialise rather than spend money (the young, the old, the ethnic minorities, those of any description who are seen to be loitering without intent to buy, the homeless in search of warmth) are simply denied access. If they succeed in entering the mall to protest (such as mall rats who 'mallinger', sitting on the floor of the mall), they are unceremoniously ejected as discordant elements.

'Gated communities' constitute a further prominent form of privatising spaces of consumption, bounded residential spaces, protected by security guards, security fences and CCTV surveillance systems.[2] These newly defined spaces of consumption typically contain a range of retail services as well as residences. Gated communities have emerged both in the inner city, sometimes linked to gentrification, in the suburbs and on the urban fringes. These bounded spaces of consumption contain relatively affluent residents who lead increasingly cocooned and privatised lifestyles, insulated from the chaos and risks of public spaces, with public interaction reduced to private (home-based) passive consumption of media images (Bauman, 2001). For those residents working for a wage, this further dichotomises daily lives between spaces of work beyond the community and a space for living, leisure and recreation within it; for those of retirement age, life becomes further confined within the walls of the community. However, residents of the community rely on casualised (typically contingent, sometimes illegal) labour provided by a new fraction of the working class, to supply consumer services (in cleaning, restaurants, leisure services) to service their lifestyles in these protected spaces of consumption. These changes in divisions of labour and practices of consumption raise questions about inter- and intra-class divisions and about the bases of individual and collective identities.

9.5 The home as a space of consumption

The dwellings in which people live can be constructed in a number of ways: as commodities, via private capital; as social housing by the state; or informally in a variety of ways by those who then live in them.[3] While a dwelling may represent a financial investment for some, a home represents an emotional investment for all, a space endowed with multiple meanings by those who live there, linked to practices of consumption (Smith, 2003). Recently, there has been renewed interest in the home and the domestic sphere as spaces of consumption. In the nineteenth century in the USA, 'commodification of the home provided seemingly endless possibilities [for capital]. By the end of the nineteenth century, food, shelter, clothing and home furnishings had all become commodities' (Domosh, 1996, 260). Similar processes then emerged in the UK, emergent capitalist economies in continental Europe, and then in other parts of the world. By the start of the twenty-first century, the range of commodities present in homes in the late modern world had expanded enormously, from diverse consumer durables and 'white goods' to cut flowers (Hughes, 2000).

FIGURE 9.2 A new residential space: St. Peters Basin, Newcastle upon Tyne, England

Consumption in the home has both symbolic and material dimensions. Consequently, 'each time we bring a product home our home is transformed physically – something is there which was not there before – and transformed in terms of values and meaning' (Sack, 1992, 3). Sack specifically refers to furnishings and the ways in which these artefacts are deployed in creating particular environments and spaces, such as living rooms and drawing rooms. Such rooms are both material and discursive constructions, embodying and expressing the values of the people who make and perform them, as home making has become very closely related to identity formation (Lofgren, 1990). The emphasis on these aspects of 'home making' has been linked to the rise of specialist magazines, and also to new and re-vamped spaces of sale such as Ikea and Habitat. Such 'lifestyle' stores typically produce their own magazines and catalogues. In addition, new magazines such as *Wallpaper* and *Elle Decoration* have redefined ideas of acceptable décor and styles of home furnishing. As a result, 'the home has become an important arena of playfulness and creativity ... the project of "home" is never complete and through constant visits to furniture stores such as Ikea, notions of "family" and "gender" are constantly renovated'. This emphasises the performative character of 'home' and its links to identity (re)formation but

one can proceed too far in celebrating consumer autonomy, reflexivity and resistance. Producers and marketers exploit consumer interests in playing with identity in order to speed up fashion cycles in furniture and reduce its durability.

> A consideration of the space of the home reveals the complex process which transpires when goods move from the store to the home and are situated within the home. (Leslie and Reimer, 1999, 414)

This is an important qualification, linked to the creation of meanings via advertising and the imperatives of capital accumulation as expressed in the pressures constantly to product innovate.

On the other hand, there are pressures to subvert the logic of mainstream markets and the intended meaning of advertisers via practising alternative styles of home making. One example of this is the purchase of second-hand furniture, white goods and other household goods from car boot sales, or from spaces of sale in the social economy, sometimes selling re-conditioned items. The movement towards low-cost home renovation and re-decoration is both reflected in, and given further impetus by, the rise of TV makeover programmes such as, in the UK, *Changing Rooms* and *Groundforce* (extending from interior to exterior home making). This can be seen as involving the process of (re)making homes as a source of creativity, reflexivity and transformation, as well as sites of performance and of re-making social (especially gender) relations that challenges the logic of the mainstream economy and promotes the home as a space of alterity.

Food is another important arena of consumption in the home (Crewe, 2000, 279). As well as enabling the physiological reproduction of the body, there are symbolic dimensions to food consumption within the home and the privileging of the family as a site of consumption, sitting around the table to eat together as a collective familial act. The preparation and consumption of food and its cultural meanings are embedded within webs of household power relations, constantly (re)producing gender divisions of labour as part of an economics of domestic consumption. The rich meanings around food and family reveal much about contemporary articulations of belonging and subjectivity, with the result that 'food and eating may serve to embody and render fleshy the neat abstraction of citizen' (Probyn, 1998, 161). Privileging the family as a space of consumption consolidates the basis for the 'familial citizen'.

Miller (1991) examines food in the context of the creation of inter-war suburban USA as a specific sort of consumption space. Suburbanisation involved creating new residential living spaces, containing a self-service society dependent upon household appliances produced and purchased as commodities. Magazine advertisements, promoting the growing diffusion and adoption of 'white goods' – vacuum cleaners, refrigerators and other household appliances – were central to the construction of this new consumption space. The sale of these commodities was promoted in a very gender-specific way – 'Madam, you need never sweep or dust again' – allied to allusions to (or maybe illusions about?) the rationality of science and a modernist belief in scientific progress. This was projected as providing the basis for new forms of household management, replacing domestic labour by new consumer durables, outputs of the new mass production 'growth industries' of a burgeoning Fordism. Purchasing such devices would lead to cleaner homes, reducing the risks of ill-health because of improved standards of food hygiene: 'You, as a conscientious mother, buy the best food for your children, prepare it with scrupulous care and cook it correctly. Yet, in spite of all, you may be giving your children food which is not wholesome – and possibly dangerous' (cited in Jackson and Taylor, 1996, 358). Even so, Miller's account tends to assume that the message would be read 'as intended' by those propagating it rather than its recipients challenging and subverting the intended meanings, creating alternatives to them. As such, it underplays the potential agency of consumers as active, knowledgeable subjects.[4]

In summary, the home is (performed as) a space of manifold forms of consumption: flowers, food, furniture, 'white goods' of various sorts, and so on. Most of these are purchased as commodities, although not necessarily by those consuming them. Consumers spend considerable time cleaning, discussing, comparing, reflecting, showing off and even photographing many of their new possessions (McCracken, 1988). In general, however, relatively little is known about what people actually do with those commodities and other things that they buy, on how they transform them through repair, restoration or alteration, or how they display or dispose of their possessions via re-sale or giving gifts (Crewe, 2001, 280). Clearly, there is some way to go before providing a satisfactory response to the challenge 'to conceptualise consumption as an ongoing process'.

9.6 The body as a site of consumption and the construction of identities

Recently, there has been considerable emphasis upon the emergence of 'reflexive consumption' grounded in cognitive reflexivity (Giddens, 1991), underpinned by abstract systems of knowledge, 'expert systems', that allow individuals to construct, monitor and regularly reconstruct the self and its identities. In this era of 'reflexive consumption', issues of individual identity construction, grounded in aesthetic as well as cognitive reflexivity, have become central. This reflects more than the proliferation of styles associated with Bourdieu's (1984) notions of invidious social 'distinction'. Of greater importance is the process of 'the decline of tradition which opens up a process of "individuation" in which structures such as the family, corporate groups and even social class location, no longer determine consumption decisions for individuals'. Swathes of lifestyle and consumer choice are liberated and individuals forced to take risks, to bear responsibilities, and to be actively involved in the construction of their own identities, to be enterprising consumers. These processes are largely responsible for the shift to 'the semiotization of consumption, whose increasingly symbolic nature is ever more involved in the self-construction of identity' (Lash and Urry, 1994, 61). These are bold claims, strong assertions.[5]

Such claims have validity for those social groups able to exercise choice in the market place, especially the new middle classes of late modernity who become what they eat, what they wear, where they go, and so on. In this sense, commodification extends capitalist mechanisms into the previously forbidden territory of the body, as subject positions are created and inflected by particular consumption practices and spaces which act as 'binding agents' (Thrift, 2000). However, in other respects, these are contentious claims. There are enormous cultural and economic inequalities in access to, and ability to use, abstract knowledge and 'expert systems'. Lash and Urry (1994, 231) partly recognise this, emphasising the fundamentally aesthetic nature of interpretative dimensions of contemporary everyday life. A proliferation of images resulting from 'the expansion of cultural or culturally attuned systems' means that 'aesthetic cultural capital is now important in the everyday lives of most western people'. As such, the active construction of individual identities has become more important, grounded in expanded aesthetic and cognitive reflexivities and supported by enhanced cultural and social apparatuses of reflexivity (Glennie and Thrift, 1996b, 232). Consequently, 'each person's biography is removed from given determinations and placed in his or her own hands; open and dependent on decisions'. The proportion of life opportunities that are fundamentally closed to decision making is decreasing and the proportion of the biography that is open and must be constituted personally is increasing. 'Individualization of life situations and processes means that biographies become self-reflexive. Socially prescribed biography is transformed into biography that is self-produced and continues to be produced' (Beck, 1992, 135).

However, the reference to 'the everyday lives of most western people' implicitly acknowledges the great socio-spatial differences between the core and peripheral territories of the capitalist economy. Consider, for example, the marked inequalities in access to the Internet and illiteracy rates in many parts of the world. Not least, 'non-western' people are by far the great majority of the world's population, while there is great socio-spatial differentiation within the broad territories of both core and periphery. As such, claims that 'the late twentieth century [is] a particularly flexible period in the development of novel consumption practices, because sociability has become more autonomous of other social relations ... and because of the expanded scope and scale of reflexivity' (Glennie and Thrift, 1996b, 236) require careful qualification. Consumption decisions and the lifestyle trajectories of individuals were never wholly determined by structures, amenable to being 'read off' from structural position. People are not simply passive dupes, even in the most dire of circumstances. However, individuals undeniably vary in their capacity to exercise such self-determination because of their structural (class, ethnicity, gender, place and so on) positions. Claims about self-produced reflexive biographies have a hollow ring in the context of people such as those in sub-Saharan Africa where the main priority for most is to secure enough calories to survive.

Advertising undoubtedly seeks to convince people that they can construct unique identities via commodity consumption, especially in the core territories of the capitalist world. Many choices about self and biography are made through markets. Consumer goods have increasingly become, for some, the medium through which gender and sexual identities are sought out and fixed, and they have been used in increasingly creative and flexible ways. For example, 'particularly since the mid-1980s there are numerous instances where industrial production and advertising has followed in the wake of youthful tribes within the throng rather than vice versa' (Glennie and Thrift, 1996b, 234). Market segmentation, fragmenting previously created mass markets, based upon defining particular types of consumer has been central to advertising and marketing strategies for several decades. Such segmentation has more recently been enhanced via creating 'sub-brands', brand niches to exploit an enhanced aesthetic meaning, for which more affluent customers will pay premium prices to acquire the cachet of identity that they confer. Increasingly individualised 'specialised consumption' of niche marketed cultural and other commodities has reinforced tendencies to create 'disembedded lifestyle enclaves' (Lash and Urry, 1994, 142).

Often this involves projecting particular bodily images as the quintessence of a particular identity or lifestyle, encouraging people who aspire to such identities to consume in particular ways. Bodily appearance has long been significant in relation to employment in some occupations, and more recently has assumed greater importance in others. As such, it is important in the allocation of jobs via the labour market. McRobbie (2002, 100) draws attention to the 'culture of "pampering yourself" with beauty treatments, health farms and any number of body-toning therapies', particularly important in the new service sector which 'now expects the workforce to look especially attractive for ... aesthetic labour'. Bodies thus become new sites of consumption as a necessary condition for being allowed to perform particular sorts of embodied labour. However, bodily appearance is linked more widely to issues of identity and meaning. The body is always inescapably encoded by cultural norms (Negrin, 1999). There is clear evidence that 'you are what you wear', that 'you are what you choose to adorn you body with' in terms of cosmetics, perfumes and so on. This is intimately linked to issues of individual and, in some respects, collective identities, and to issues of citizenship and political action (Crewe, 2001, 632).

The associations between consumption and identity formation vary over time/space, as can be illustrated with respect to clothing. In the nineteenth century in the USA merchants invested heavily in advertising to convince middle-class women that factory produced ready-to-wear clothing was better, cheaper and 'simply more modern' than home-made clothing. Adverts played on the woman's sense of duty 'to provide the best up-to-date care for her family, as well as 'the status that was associated with the most stylish clothing'. Such appropriate patterns of commodity purchase and consumption enabled a middle-class woman to ensure that her appearance, and that of her family, 'acted as important indicators of social status' while 'correct consumption also served as reflections of her role as wife and mother' (Domosh, 1996, 260–1). In due course, the consumption norms of the middle classes cascaded down to the working classes, seeking to emulate the former, who in turn sought out new forms of commodity consumption.

In contrast, a century later emulation no longer has the same influence. Dress styles in late modernity are much more personality-specific than specific to social positions. Thus they involve an important set of identity choices and identity risks, especially for young people. These often involve creating multiple identities. Thus, for example, through dressing up, consuming clothing and presenting the body in particular ways, through wearing and using appropriate consumer goods, often as part of tribes within the throng, new identities can be identified, experimented with, and explored (Glennie and Thrift, 1996b, 236). More generally, 'playing' with commodities and experimenting with new styles and identities requires a social setting and the reactions of others to guide choices. However, there is great socio-spatial selectivity in terms of who can participate in such explorations of the self. While it may be possible for more affluent people in more affluent parts of the world, it is debatable as to its validity beyond those bounds.

Food consumption may also be critical in relation to identity formation. Referring to decisions as to what to eat, Whatmore (1995, 36) suggests that 'the prevalent representation of such experiences as the mark of "consumer choice" belies a diminished understanding of and control over what it is we are eating and the social conditions under which it is produced'. There is, therefore an 'intimate and unavoidable connection' between the food system, retailing and the consuming body (Crewe, 2001, 630), which is at best partially understood by many consumers of food products. Concerns over these connections, and of public understanding of them, has mushroomed in the wake of growing fears about food quality, registered in increasing worries as to food consumption. Consequently, food consumption has become more risky and more reflexive, with a growth in 'more careful consumption' among certain sets of consumers (Marsden, 1998, 285). This has translated into an expansion in quality assurance schemes as these concerns are transmitted down the food chain, driven by major food retailers in order to meet requirements for due diligence. However, because of the specific character of the food production system and its grounding in bio-chemical processes, it is difficult to ensure quality. Not least, this is because nature evades the outflanking processes of industrial capital, 'bouncing back in the wake of human modification'. The most notable example of nature's 'boomerang quality' is BSE, 'where a seeming domestication of various natural entities gave rise to a terrifying new actor (a prion protein), one that causes irreversible damage to the brain' (Murdoch et al., 2000, 110). This generated growing concerns about the consumption of certain sorts of meat (notably beef), fuelled by fears about the bodily effects that this might produce via the trans-species transfer of BSE to humans in the form of variant CJD.

However, there are sharp social divides in relation to these growing concerns about food quality, a result of varying knowledge and endowments of cultural capital and differential incomes and discretionary purchasing power. Such concerns are most strongly expressed in the more affluent parts and social strata of the late modern world. For 'while public consciousness has been raised by food scares, this shift towards a logic of quality is fuelled by the emergence of a growing food elite who are knowledgeable about tracing the origins of their foodstuffs' (Murdoch et al., 2000, 110). As such, this process is deeply divisive. The emergence of politically, socially and environmentally aware and savvy consumers belies the 'gritty reality that the majority of British consumers have neither the political clout nor the financial means to engage in careful consumption and mobilize against the dictates of big retail capital' (Crewe, 2001, 631). If this is the case in Britain, then it is true in spades over much of the world where the prime concerns are lack food of any sort and the immanent threat of starvation rather than of food quality (Seager, 1995).

While we may be 'what we eat', it is also evident that people – especially women but increasingly men – are exhorted by a constant stream of adverts to work on their bodily appearance and presentation in other ways. These include working out at the gym, using particular cosmetics and wearing particular styles and brands of clothing. Showalter (2001, 3) suggests that retro or vintage is the ideal feminist choice because it was 'an ironic style that inserts a wearer into a complex network of cultural and historical reference'. This is but part of the story, however. Commodification of the body through the fashion and beauty industries 'pre-supposes that acutely self-conscious relation to the body which is attributed to femininity'. The effective operation of the commodity system 'requires the breakdown of the body into parts – nails, air, skin, breath – each one of which can be constantly improved through the purchase of a commodity' (Doane, 1987, 32). In the twenty-first century, however, this is as relevant to certain versions of masculinity as to femininity. Concern with bodily appearance has become much less gender-specific, at least in some parts of the world and among certain social strata.

In brief, 'fashion – and its connection to food, the body and gendered subjectivity – is finally being taken seriously' (Crewe, 2001, 634). One consequence of this is that 'eating disorders' such as anorexia nervosa and bulimia are being radically re-evaluated, understood not as individual pathologies but as culturally, politically and discursively located forms of embodiment and body management. Food is increasingly considered 'not in terms of its relevance to eating disorders, excessive thinness or obesity, but the fashion and food connection is also considered in terms of fetishism, seduction, the spectacularity of the catwalk show and contemporary concepts of the grotesque' (Crewe, 2001, 634). This raises important questions as to the relationships between the culturally coded normative images of bodies projected via advertisements as ideal and desirable and the impacts that these have on bodily appearance and well-being. While some may contest and seek to subvert such adverts, many others clearly accept them at face value and seek to manage and present their bodies accordingly, irrespective of the mental and physical damage that this may cause. Furthermore, addictive and excessive consumption may have other adverse impacts on health, well-being and lifestyle, leading to debt, imprisonment and sometimes suicide.

For others their bodies become things to be commodified, for example selling organs to others as a desperate response to poverty.

In summary, the increasing penetration of commodity relations into and commodification of the lifeworld and the accelerating circulation of images and objects of consumption has further highlighted relations between consumption and identity. However, there are alternative interpretations of these processes (Glennie and Thrift, 1996b, 221–2). The first, grounded in a dystopic notion of post-modern economies and societies, suggests that this increased speed of circulation empties both objects and subjects of meanings, creating cultural fragmentation in the sphere of consumption as formerly stable social markers, such as class, lose their relevance. The alternative emphasises the possibilities that speed up creates for reflexive modernisation, and for redefining and reconstituting (*inter alia*) the meanings of people, spaces and things via consumption, culturally segmenting consumers, and creating a myriad of sub-groups of consumers in place of earlier and larger class-centred constellations.[6] While consumption has always been determined by more than simply class, class relations also continue to influence patterns and spaces of consumption in capitalist societies. Segmentation of consumer markets continues to be heavily circumscribed by purchasing power and 'relations of consumption are fields of power struggle and reproduction' (Slater, 2002, 74). However, both interpretations neglect aesthetic issues that have reinforced links between consumption and identity. Aesthetic reflexivity, grounded in hermeneutic knowledge, allows people as subjects in the sphere of consumption to construct their own identities through engagement with consumer and lifestyle choices (Allen, 2002, 41).

However, this raises key issues as to which people, in which spaces, can exercise such agency in the socio-spatially-specific process of constructing their identities. While 'fractions of the middle class have become more educated into a controlled de-control of emotions and the sensibilities and tastes that support a greater appreciation of the aestheticization of everyday life' (Featherstone, 1991, 81), they constitute a small minority. Overall, even in the core territories of late modernity, 'relatively few' people take consumption decisions 'with aesthetic considerations in mind'. In contrast, 'the rationalisation of daily life is the dominant tendency emanating from economic activity, with ramifications in the fields of domestic and family life, leisure and recreation, as well as ordinary consumption' (Warde, 2002, 198–9). For the vast majority of the population everyday life remains profoundly functional and instrumental rather than aesthetic in its orientation.

9.7 Spaces of consumption, spaces of work

Consumption of spaces requires work by their consumers while consumption in these spaces requires the work of co-present producers and consumers of simultaneously produced and consumed commodified services (Walker, 1985).

Alternatively, consumption may require unpaid work by consumers and/or providers of services themselves – for example, in home making and various DIY activities, or in emotional and personal caring work in the home or elsewhere. The nature of consumption work in the home varies over time and space. For example, the nineteenth-century commodification of the home in the USA transformed the work of middle- and upper-class women from domestic production to public consumption (Domosh, 1996, 260). Similar transformations later took place as commodification penetrated homes elsewhere. In the spaces of late modernity, there are socio-spatially specific processes of 'pseudo-commodification' of domestic work, with payment from some household/family members to others. Children are paid to carry out household tasks, work formerly done on an unpaid basis on the basis of obligation, respect and love.[7]

Service work within the social relations of capital outside the home involves a variety of forms of paid employment that encompass affective, cultural and economic dimensions. It comprises a 'contingent assemblage of practices built up from parts that are economic and non-economic (but always cultural) and forged together in pursuit of increased sales and competitive advantage' (Du Gay and Pryke, 2002, 4). Companies must therefore maintain the 'frontier of control' and manage the labour process in ways that reflect the co-presence of service provider and consumer, the necessity to cede considerable autonomy to workers, and acknowledge that the appearance and performance of the provider is often experienced as part of the service. In many occupations in service-based economies, from fast food to fast money, the service or product has become inseparable from the process of providing it (McDowell, 1997, 2001). Such dialogic co-production of services in a context of increasingly individualised consumption can lead to fine classificatory distinctions, especially where consumers are members of the new middle classes (Bourdieu, 1984).

The co-production of services requires that workers with specific social attributes, from class and gender to weight and demeanour, be disciplined to produce an embodied performance that conforms to idealised notions of the appropriate servicer. In this normalisation, the culture of organisations, their explicit and implicit rules of conduct, becomes increasingly important in ensuring that workers have desirable embodied attributes and in establishing the values and norms of organisational practices. Thus the service is often intimately connected to the identity and performance of the person providing it. Although recently emphasised in relation to some services, it has long been the case in relation to others. For example, the organisation of banking in the UK has 'always been tied into the embodied characteristics of gender, class, age and race, and to the notions of appropriate behaviour and style' (Halford and Savage, 1997, 116). This was linked to specific paternalistic recruitment practices and spaces of recruitment. Such practices were perhaps most prominently visible in the financial service nexus of the City of London, controlled by a tight group defined by education (and socialisation) at particular public schools and Oxbridge, and appropriate accents and tailors. There is, however, evidence that even these are changing because of the demands of new methods of working and social relationships of production (Allen et al., 1998, 92) and that the precise mix of desired characteristics has altered along with banking practices (McDowell, 1997).

The imperative for workers increasingly to conform to the highly prescribed, almost theatrical, requirements of 'body work', now extends to working in a diverse range of interactive service sector activities, including fast-food outlets, restaurants, security work, and tourist and entertainment facilities. As in banking, this powerfully shapes recruitment and retention criteria and practices (Allen et al., 1998, 103–4; Crang, 1994; Urry, 1990). Typically workers are chosen because they possess appropriate sorts of cultural capital in terms of age, gender, bodily appearance, weight, bodily hygiene, dress and style, and interpersonal skills in interacting with customers, which combine to produce an acceptable workplace persona. However, although much of this work is poorly paid and often precarious, it may nevertheless provide satisfying work for those who perform it precisely because it is based on interpersonal interaction and co-production of services. For example, poorly paid workers in restaurants have considerable autonomy in organising their work activities, much of which consists of social-ising with customers who are also often friends (Marshall, 1986). To a consid-erable extent, workers performatively interacting with customers become the product. Reflecting this, remuneration, while low, is often performance-related, linked to individualised ways of organising work and assessing performance. Wage levels reflect individual productivity and additional payment in the form of tips reflect the consumer's satisfaction with the service provided. However, in so far as workers become the products, they may become involved in particular forms of self-exploitation. This is a *fortiori* the case in (typically illegal) services such as prostitution. Moreover, a considerable amount of interactive work 'is so thoroughly imbued with an economising logic that it begins to undermine the conventions of face-to-face interactions'. Consequently, 'the extent to which ser-vice industries are delivering face-to-face and intangible products has been exag-gerated and ... the labour processes are less distinctive from other kinds of work than is usually implied' (Warde, 2002, 190).

However, the availability of appropriate labour can influence the location of spaces of service production, especially when this involves combinations of personal attributes and social and technical skills. For example, in many financial and technical service sector occupations know-how and 'knowing how to go on' is critical. Success depends upon combining 'soft skills' with 'hard technologies' to deliver a particular service, emphasising the importance of employees with suitable skills and attributes. On the other hand, the production of many services is labour-intensive and so companies seek to control labour costs without prejudicing the quality of service delivery. In part, this has involved relocating service activities not requiring face-to-face interaction, or those in which distan-ciated contact via telephone or the Internet can substitute for face-to face inter-action, to new locations in which cheaper but appropriately qualified and skilled labour is available. These activities require temporal, but not spatial co-presence for service provision. In part, it has involved 'disintermediated' service provision, substituting machines for people as the immediate point of contact. The proliferation of call centres, websites and portals is indicative of these tendencies, as new spatial divisions of labour in service production have been constructed in response to the possibilities offered by advances in ICTs (Lakka, 1994). However, this distanciation of provision creates requirements for new forms of regulation of work and of ensuring quality of service provision, for example, via state policies to make sure that suitably qualified labour continues to be available and that ICT infrastructure is of high quality.

In circumstances where such opportunities are unavailable, companies seek other ways to control labour costs and manage workers. Subject to regulatory limitation and the specificities of local labour market conditions, companies may replace permanent with temporary or casual workers. This has occurred in a variety of service activities in which the level of labour demand fluctuates markedly over time and in which there is considerable demand for large amounts of unskilled labour (such as contract cleaning or simple office work). In services such as banking, employers cut labour costs by substituting machines (such as ATMs and telephone banking) for people and changing the terms, conditions and organisation of work for those employed. New forms of functional and numerical flexibility, involving tiered recruitment strategies and increased use of casual and part-time female clerical staff, allow companies to match labour supply more closely to demand within deeply segmented internal labour markets (Halford and Savage, 1997, 112). These tendencies have been facilitated by the growth of temporary employment agencies, which deliberately help create the labour market conditions that make such recruitment strategies possible (Peck and Theodore, 1997). More generally, casual recruitment practices redolent of activities such as dock work are being re-invented in parts of the unskilled and deskilled service sector of contemporary capitalism, as spaces of service provision become sites of increasingly precarious employment for those employed in them.

Taylorist principles have increasingly been introduced into large swathes of routine service sector activities to control labour and labour costs. McDonald's restaurants epitomise a well-established tendency in the fast-food sector for mass production of standardised meals and for carefully scripting standardised forms of work that prescribe precisely what employees are to do and say. In 1948 McDonald's 'divided the production of their restaurant food into discrete (skill-less) tasks applying the time and motion paradigm that had ruled factory assembly lines. Machines were and still are designed so that workers could reach high speed in minutes' (Westwood, 2001, 11). The aim is to make the production and delivery of a given meal identical in all McDonald's restaurants (Leidner, 1993), with kitchens 'full of buzzers and flashing lights that tell employees what to do', with cooking instructions often designed into the machines (Westwood, 2001, 10). Such process of Taylorisation are also marked in service sector activities that involve processing large amounts of data and/or paper or dealing with customer enquiries, leading some commentators to refer to the 'industrialisation of white-collar work' (McRae, 1997). Indeed, in white-collar industries such as advertising, the labour process was 'regularised and Taylorized' from the early part of the twentieth century (Lash and Urry, 1994, 139). Finally, Taylorist principles are in some ways even penetrating the provision of medical services, with the deskilling of operations such as those for cataracts, now performed *en masse* by medical technicians rather than doctors, made possible by process innovations in operating technologies (Metcalfe and James, 2000).

New forms of work organisation often also involve reworking and extending Taylorist principles in spaces of service production. In recent years there has been an increased 'contractualisation' of employment, with people employed on different contractual terms in respect of hours, benefits and entitlement. The generalisation of employment insecurity thus becomes a regularised feature of working life. While often described as 'flexibility it is therefore perhaps more accurate to refer to this "individualisation" of employment relations as a form of Taylorism' (Allen and Henry, 1997, 185). This new Taylorism of employment relations reflects the temporal limitation, legal (non)-protection and contractual pluralisation of the employment of labour (Beck, 1992, 147). Thus the growth of contingent and (sub-)contract employment represents further widening and deepening of social and technical divisions of labour as flexibility secures, for employers, more efficient and possibly cheaper ways of doing things. However, employees experience such flexibility as risk and uncertainty.

9.8 Summary and conclusions

Consumption is a critical moment in geographies of economies. It marks a transition from emphasis upon the exchange value characteristics of the commodity to those of its use value. Consumption occurs in a variety of spaces, both public and private, formal, informal and (on occasion) illegal, and is often closely tied to the creation of individuals' identities. Thus consumption has affective, emotional and symbolic dimensions, as well as material and economic ones. Performing consumption in various ways becomes one route through which people, especially relatively affluent people in affluent spaces, seek affirmation and confirmation of the identities that they seek to create. Miller (2002, 182) seeks to reconnect debates about consumption with a broader systemic understanding of the political economy of capitalism. He suggests that 'ordinary mundane consumption' is neither hedonistic, materialistic, nor individualistic. Instead, it is above all the form in which capitalism is negated and 'through which labour brought its products back into the creation of its humanity'. This progressive development of consumption is 'the background to the rise of virtualism'. Virtualist institutions have developed 'through a sleight of hand that took the authority of the consumer and appropriated it for the interests of the institution in general and to the detriment of the people in their social role as consumers (and in most cases also as workers)'. Miller concludes: 'If consumption has developed as the negation of capitalism until it was trumped by virtualism, then virtualism could be argued to be the negation of a negation with a grand narrative.' However, many people in peripheral spaces in the capitalist economy lack the economic capacity to perform even 'ordinary consumption'. Crucially, they lack the money needed to purchase most, if not all commodities, not least as they cannot sell their labour-power on the market. For them, the issue is one of day-to-day survival via a variety of non-commodified subsistence strategies rather than the negation of capitalism via commodity consumption.

Notes

1 'De-differentiation' as a social practice clearly anticipated the introduction of the term to late twentieth-century post-modern social science.

2 Some 14% of the population of the USA live in such communities (Minton, 2002).

3 In the cities of the developing world, between 30% and 70% of people live in 'informal' settlements and dwellings (United Nations-Habitat, 2003).

4 Other studies of consumers' reactions to adverts are more sensitive to this possibility, consistent with a recursive approach to circuits of meaning (see also Chapter 4).

5 Moreover, Lash and Urry (1994, 108–9) claim that these processes underlie the shift to small batch production of commodities and the proliferation of advanced consumer services, which provide professional help (and 'expert systems') to de-traditionalised individuals. This version of the myth of consumer sovereignty ignores powerful systemic imperatives to devise new methods of HVFP and mass customisation of commodities (Chapter 7) and the advertising and marketing strategies of companies producing them (Chapter 4).

6 There are parallel arguments concerning the weakening of former work-based identities and the selective re-creation of new ones (see Chapters 7 and 8).

7 For example, one-third of parents in the UK paid their children for housework, some paying up to £1,200 per annum (Brun-Rovet, 2002).

10 From Spaces of Pollution and Waste to Sustainable Spaces?

10.1 Introduction

Economic activities necessarily involve chemical and physical material transformations. As such, they have unavoidable and often unintended and unwanted effects on nature and natural eco-systems (in so far as any eco-system can be so described in the face of pervasive human impacts). Consequently, economic processes result in varied forms of environmental pollution. Pollutants can be defined as 'xenobiotic substances and natural substances in unnatural concentrations' (Weaver et al., 2000, 37). Pollutants and wastes have replaced neo-Malthusian fears of resource depletion posing 'limits to growth' as *the* critical environmental problem (Young, 1992, 5). Nevertheless, the natural environment continues to perform as a sink of infinite capacity for free deposition of unwanted wastes, the unpriced by-products of commodity production. For example, bulk chemicals production can produce 5kg of unwanted by-product for each kilogram of intended output; this ratio can rise to 50:1 for fine chemicals (Luseby, 1998). Unless the ecological problems posed by mass pollution and waste generation are adequately addressed, the problem of resource exhaustion may never arise.

10.2 Necessarily localised spaces of pollution and waste

Although many economic activities have considerable locational room for manoeuvre, others do not. Mining (and often associated primary processing) remains unavoidably fixed, tied to the location of natural materials because of accidents of historical geology. Agriculture, forestry and fishing are strongly constrained by variations in natural ecologies, despite efforts to emancipate them from such constraints. As a result, such activities often generate a range of localised environmental impacts that impinge upon the health and living conditions of local residents. These can be experienced as noise, visual intrusion, and pollution of the atmosphere, land and water bodies by hazardous and noxious materials (Beynon et al., 2000). While such impacts can be ameliorated by technological fixes, they cannot be eliminated. More generally, the unwanted environmental consequences of economic growth and industrialisation remain heavily localised, between and within countries. The harsh reality of pollution is time/space specific, impacting upon spaces in which people live and work, on their health, and on death rates. People live in visually blighted landscapes, punctuated by the noise of mining, processing and manufacturing, and breathing polluted air. There is, however, often a distance decay effect from the pollutant source, creating an externality gradient relating intensity of impacts to distance from source.

Such impacts have been endemic throughout the course of industrial capitalism. Environmental costs were accepted and interpreted as a necessary price of employment and growth. Attempts to preserve or protect the natural environment often provoked powerful alliances (of companies, trade unions and local political organisations, for example) seeking to prioritise employment, production and profits over environmental concerns (Blowers, 1984). Such conditions have been ameliorated but not eliminated by state regulation in much of the advanced capitalist world. The initial state policy response there was to seek to manage and contain the environmental impacts of industrial production. This involved land use planning, segregating incompatible land uses, along with public health legislation and some weak environmental regulation to limit the worst excesses of such impacts. Environmental pollution was seen essentially as a *localised* problem. However, industrial activities continue to have localised environmental impacts, some unavoidable if these activities are to continue. They could be reduced but not entirely removed by using more appropriate technologies, precisely because they are processes of material transformation subject to the iron laws of thermodynamics.

As jobless and then job-shedding growth emerged in such industries, protests grew in the surrounding localities against the impacts of polluting production. Claims that a trade-off between employment and the environment was inevitable, that industrial pollution was an unavoidable condition of everyday life, no longer had credibility. Increasingly, existing forms of state regulation appeared inadequate to an emerging 'green' politics. Growing environmental concern led to pressures to tighten environmental standards. People living in towns and villages built around industries such as coal mining, chemicals and steel production were increasingly resistant to their adverse polluting effects. However, such environmentally-based politics and concerns were themselves unevenly developed and environmentalist movements to curb industrial pollution were often fiercely resisted (Beynon et al., 1994, 2000).

The growth of complex urban industrial economies continues to have severe localised effects. In many parts of the world, processes of urbanisation and the resultant expansion of the built environment are leading to the further

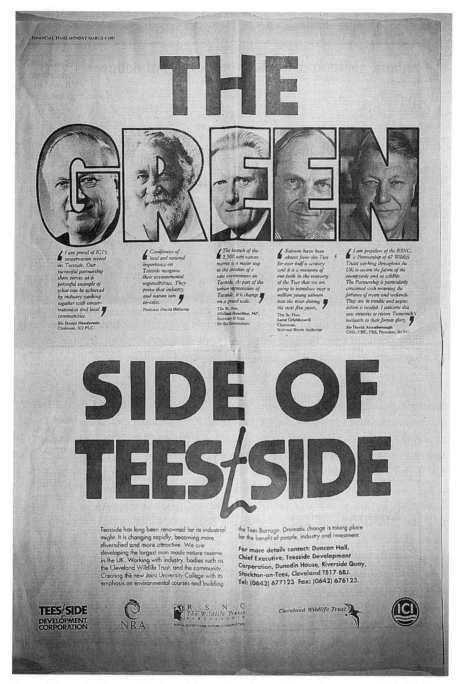

FIGURE 10.1 From a pulluted space to a sustainable space? The Green side of Teesside, England

FIGURE 10.2   From a polluted space to a sustainable space? Reclaiming the
site of Vane Tempest colliery for residential use, Seaham, England

substantial conversion of rural to urban land and the destruction of natural
environments and ecologies. These processes are currently perhaps most sharply
expressed in the explosion of massive urban areas over much of South America
and south-east Asia but are far from confined to there. For example, between
1980 and 1990 almost 14% of land 'previously considered to be part of natural
cover was lost to urban development and housing' in 11 countries of the
European Union (Commission of the European Communities, 2001, 26). In
addition, the amount of land converted to road space increased by 11% between
1980 and 1998.

Modern 'industrial' agricultural practices can cause severe localised pollution (for example, via production of organic wastes) and also generate more
widespread unintended environmental effects, such as increased soil erosion and
silting of rivers and fertiliser pollution of inland waters. Nitrates can no longer
be held down by the colloids of vegetal soil and are carried away by running
water to accumulate in coastal waters and lakes (Deléage, 1994, 40). Even
more seriously, modern agriculture can involve deliberate wholesale ecological
destruction. For example, clearing equatorial rainforest to provide cattle ranches
markedly reduces genetic variety, bio-diversity and species diversity (Yearley,
1995a). The result can be extensive spaces of waste. This tendency has recently
been given a further twist. By permitting the isolation and removal of value-
producing and apparently multipliable genes, allied to their preservation in

genebanks in the USA and Europe, genetic engineering removes the incentive to preserve the plants, their communities and ecologies of which those genes are a constitutive element. Such areas run the risk of becoming biological wastes, depleted of their natural biotic communities.

10.3   Locational choice and the selective siting of polluting industries

Whereas the pollutant effects of some activities are necessarily localised, companies engaged in other activities have greater freedom of locational choice. As such, they can site polluting industries in spaces of least resistance – from parts of the global peripheries to peripheral regions and inner cities in the core areas of capitalism. For example, Chicago's south-east side is plagued by numerous pollutant industries, commercial hazardous waste landfills and toxic waste dumps. Consequently, it has one of the highest rates of incidence of cancer in the USA (Bullard, 1994, 279–80). In Sydney, Nova Scotia, the tar ponds, the legacy of the now-closed steelworks, sit in the middle of the built-up area, surrounded by signs warning people to keep out. The site – 'the most polluted site in north America' (Lotz, 1998, 167) – is extremely hazardous because of high concentrations of polychlorinated biphenols (PCBs). Times Beach, Missouri, was irreparably contaminated by dioxin, a by-product of manufacturing antiseptic hexachlorophene. The town was subsequently evacuated via a mass buy-out of its inhabitants by the Environmental Protection Agency and surrounded by fences and armed guards, with huge signs proclaiming 'Hazardous Waste Site' (Calton, 1989). In summary, such events are comparatively rare but not unknown in the internal peripheries of the core territories of capitalism.[1]

In sharp contrast, such events are only too common over much of the industrialising periphery, as ruling elites and national states unashamedly prioritise economic growth and employment over the natural environment and human health and living conditions. Consequently, companies can deploy dangerous working practices and polluting production technologies that are inadmissible within workplaces elsewhere. For example, in December 1984 five tons of poisonous methyl isocyanate gas leaked from the Union Carbide of India Ltd pesticides plant in Bhopal, killing more than 3,000 people and injuring tens of thousands. By 1999 the death toll exceeded 6,000, with substantial, and in some locations severe, pollution of land and drinking water supplies from heavy metals and organic contaminants. Residents of these areas are exposed to the risks of hazardous chemicals on a daily basis. There are many more examples of accidents at work leading to deaths, albeit on a less dramatic scale. For example, in 2003 over 30 workers were killed and 140 injured in an explosion at a fireworks factory in Wangkou, north of Beijing in China.

More generally, there is evidence of widespread problems because of the use of polluting materials and processes in many industrialising parts of the global periphery, conjuring up images of the 'dark satanic mills' of the early stages of capitalist industrialisation in the UK (Thompson, 1969). For example, Schenzen is a 'boom town' in southern China, a major centre of capital accumulation in the first decade of the twenty-first century, famous for the production of toys, Christmas ornaments and artificial trees for western markets. The health of workers is damaged through having to work with poisonous materials: 'in the toy factories the workers have to deal with poisons every day. A lot of Christmas ornaments are made with special glues and many of the factories refuse to pay for protective equipment' (August, 2002, 11). In factories producing computers and related equipment in the Pearl River delta of China, workers are often exposed to dangerous chemicals (CAFOD, 2004).

This differential capacity to pollute and produce dangerously in part reflects the increasing involvement of national states with environmental regulation, which creates opportunities and constraints for companies in their locational strategies. As a result of this, and changes in production and transportation technologies, 'dirty' industries and the production of pollutants can *to a degree* be shifted to spaces where their localised impacts are more tolerated. Historically, in the pre-capitalist era and in the early phases of industrial capitalism, industry was closely tied to the locations of heavy, weight-losing raw materials and energy sources (notably coal and iron ore) and subject to little state regulation. This reflected technically inefficient – and heavily environmentally polluting–methods of production and transport. Consequently, there was a powerful spatial concentration of industries around such materials and/or in rapidly growing urban areas. This trend continued into the twentieth century, with new consumer goods industries such as cars concentrated in big but different urban areas, spaces of production with available labour and spaces of sale for products. From around the 1960s there was growing opposition to the environmental and health effects of industrial production processes, especially those involving asbestos, arsenic trioxide, benzidine-based dyes, certain pesticides and some other carcinogenic chemicals, chemicals such as polyvinyl chloride (PVC), and some basic mineral processing activities such as copper, lead and zinc (Leonard, 1988).

As a result, environmental regulation began to tighten and companies began to relocate 'dirty', hazardous and polluting production activities, initially to peripheral regions within their home national territories but increasingly to parts of the global periphery, thereby constituting a formative moment in the creation of the 'newly industrialising countries'. Companies were often encouraged to do this by financial inducements and low (or no) levels of environmental regulation as national governments eagerly encouraged the perceived benefits of modernisation via industrial growth, regardless of environmental or social cost. As a result, the expansion of state capitalism throughout the post-colonial period has 'extended the commodification of nature and ... greatly exacerbated local causes and levels of environmental degradation' (Fitzimmons et al., 1994, 209–11). Thus the localised impacts, both positive and negative, of industrial growth have been increasingly dispersed. This added a different range of environmental impacts in such industrialising spaces to those generated by primary sector production and mining of raw materials.

The significance of lower environmental standards in relocation should not be overestimated, however. Environmental regulation certainly tightened in many advanced capitalist states from the early 1980s, acting as a push factor. For example, much of the heavily pollutant Japanese aluminium and copper smelting industries was relocated to poorer parts of south-east Asia (such as the Philippines) to avoid Japan's more rigorous environmental regulations. In Europe, there was greater relocation of productive capacity in acrylonitrile and PVC from the Federal Republic of Germany to peripheral regions in countries such as the UK and to peripheral countries such as Brazil. However, the importance of being able to dump wastes free of charge varies between industries, with relocation an uncommon corporate response to enhanced environmental standards (Leonard, 1988, 111). Most industries have responded with technological innovations, changes in raw materials, or more efficient process controls. Even when such adaptations have failed to reduce regulatory burdens, the environmental problems and the costs of responding to them have generally been insufficient to offset the attractions of location within major markets (Michalowski and Kramer, 1987). Even companies manufacturing PVC and acrylonitrile in the USA largely responded to environmental regulatory pressure and adverse publicity by technological innovation within a rapidly growing market.

10.4   Spaces of pollution, flows of wastes and the creation of new international regulatory regimes

An alternative to exporting polluting industries is to export their pollutants, either deliberately or inadvertently. For example, environmental pollutants from coal-fired power stations can be exported in molecular form via emissions from high chimney stacks, diffusing through the atmosphere and falling as acid rain and destroying vegetation hundreds of miles away. The expansion of international air and sea travel has resulted in significant emissions of pollutants into the largely unregulated global commons of the atmosphere and oceans, with adverse environmental effects (German Advisory Council on Global Change, 2002). In other cases, waste products are exported in different forms, deliberately targeting selected spaces as destinations, often on the periphery of the global economy. For example, Kassa Island, off the African coast, became the recipient of highly polluted incinerator ash from power stations in Philadelphia (Yearley, 1995b).

As people in more economically developed countries came to understand the dangers posed by noxious pollutants, and environmental standards were increased, pressures rose to find ways of coping with such pollutants. Dealing with wastes within their home territory could involve considerable financial costs. Exporting them was often cheaper than dealing with them at home, and made easier when recipient countries were misled about the nature of the wastes and/or had authoritarian non-elected governments who neither knew nor cared. Perhaps more significantly, dealing with pollutants at home could entail political costs, in the face of NIMBYism ('not in my back yard') and opposition by local communities to wastes being treated in '*their* back yard'. However, local communities have differential capacity to resist. In the UK, nuclear waste has been reprocessed at Sellafield, Cumbria, for some 50 years, with persistent worries about the effects of accidents and the exposure of workers and local residents to radiation. However, Sellafield is located in a peripheral region, with few other employment opportunities. Moreover, 'one of the best predictors of the location of toxic waste dumps in the United States is a geographical concentration of people of low income and color' (Harvey, 1996, 368). Indeed, poorer communities within the advanced capitalist world and peripheral states within the global economy have engaged in bidding wars, seeking to become destinations for hazardous wastes in return for monetary payments and incomes. Some countries are so poor, and in such desperate need of foreign currency earnings (to buy imports or repay foreign debt) that ruling elites encourage any trade likely to generate such earnings, discounting risks to the environment and the health of their populations.

Increasing environmental standards have also led to new forms of trade in wastes. Stringent regulations on recycling were introduced in Germany in the 1990s. Picking through waste to sort and recycle it is labour-intensive, poorly paid and of low social esteem. As such, it is exported to peripheral parts of the global economy – and justified as creating employment there! Although international regulatory and trade agreements have halted the worst excesses of the trade in noxious wastes, they have failed to stop it. Consequently, the global core still offloads its wastes on to the peripheries. However, there are also flows among core countries. For example, in the 1990s Japanese nuclear waste was shipped to Sellafield for reprocessing before being returned to Japan. In 2003 proposals emerged to transfer obsolete USA navy ships, replete with a variety of noxious substances (including asbestos and PCBs), to Hartlepool in north-east England for dismantling and disposal of wastes (BAN, 2003).[2] The international trade in pollutants is therefore complex. Export of wastes can be problematic for exporters, however, as the impacts of pollution return to blight their spaces of origin. For example, factories relocated from the USA into the maquiladora border zone in Mexico in response, *inter alia*, to less stringent environmental regulations there subsequently exported air pollution, sewage and contaminated food back to the USA as 'ecological havoc recognises no boundaries' (George, 1992, 6).

Thus as the locational choices of economic actors and activities widened, allowing the possibility of seeking 'spatial fixes' to deal with localised problems of pollution, there was also growing realisation that the impacts of industrial mass production and consumption were not simply spatially concentrated and confinable. Indeed, they were having global impacts, with potentially disastrous implications. This was especially so with recognition of enhanced global warming because of emissions of carbon dioxide (CO_2) and other greenhouse gases, both from fixed points and mobile means of transport. Vehicles powered by internal combustion engines generate 30% of anthropogenic CO_2 emissions (Weaver et al., 2000). Moreover, emissions of greenhouse gases from transport 'are growing more rapidly than from any other source' (Commission of the European Communities, 2001, 38). Another major global environmental problem is the thinning of the ozone layer due to emissions of chlorofluorocarbons (CFCs). In themselves CFCs are harmless to human and animal life and were widely used following their development in the middle of the twentieth century. Subsequently, they rapidly migrated to and accumulated in the upper atmosphere where, because of unanticipated chemical reactions, they destroy the ozone layer that filters out ultra-violet radiation.

As states cannot simply displace ecological problems, environmental concerns become firmly established on political agendas (Dryzek, 1994, 187). The critical insights provided by the material transformations perspective into the chemistry and physics of economic activities and processes help us to understand why this is so. Enhanced global warming threatens the sustainability of existing patterns of economic activity, production and consumption. In more apocalyptic visions, it endangers the future of the planet itself. There are serious limits to the attempts of companies and national states in the affluent core to deal with problems of industrial pollution via strict environmental regulation 'at home' and 'spatial fixes' to relocate pollution problems. For, in the final analysis, much of the pollutant effects of industrial production are simply not local and localisable. They cannot be contained through spatial fixes, only displaced to other locations from which they continue to impact upon the global environment.

Consequently, there have been attempts to construct global environmental regulatory regimes by national states agreeing to limit emissions of greenhouse gases (as at the Rio and Kyoto earth summits). At the same time, the creation of pollution trading regimes has sought to make global limits compatible with the legacies of uneven development and contemporary differences in levels of pollution production. Thus pollutants are commodified and global markets created for them by tradable permits. In this way 'business and environmental groups have agreed to extend the market to resolve ... arenas of environmental struggle' (Fitzsimmons et al., 1994, 203). However, this uneasy compromise is only an apparent resolution, displacing environmental struggles into the mechanisms of markets and the realm of economic contradictions, from whence they may be further displaced into the political arena as crises of state involvement. These tensions are further exacerbated because national states seek to influence corporate behaviour by creating international regulatory regimes but lack the capacity to enforce and monitor them. Consequently, 'their translation into behavioural effects at the corporate level is problematic' (Vogler, 1995, 154). Nevertheless, global environmental changes will impact differentially because of variations in natural environments, topography and the variable capacity to implement technological solutions to ameliorate their effects, differences that are a product of past uneven development. They will impact back upon the countries of the advanced capitalist world that generate the vast majority of such pollutants, but they have greater financial and technological resources to cope with them.

10.5   Defining and moving towards sustainability

Current economic practices of production and consumption are unsustainable for two main reasons. First, they unavoidably involve resource depletion, especially of 'stock' resources that are non-renewable over human lives. Conversely, little use is made of potentially renewable alternatives. Secondly, they unavoidably create a range of pollutants that threaten human lives over varying time scales. In the contemporary capitalist economy, the most common way to deal with unwanted pollutants involves 'end-of-pipe' solutions to ameliorate the effects of polluting technologies, allowing wastes to be captured and treated prior to release into the environment. This is the case, for example, in relation to the treatment of polluted water, pollutant gases from power stations or pollutants from internal combustion engines. However, there are serious limits to such approaches: 'they neither reduce the draw on non-renewable resources nor the overall quantity of wastes arising'. On the contrary, 'often, they actually increase both, as more raw materials have to be brought into the economy to build extra equipment and operate additional processes. ... Above all end-of-pipe responses have no effect on the quantity or type of raw materials used and so will not deliver long-term sustainability' (Weaver et al., 2000, 210). Any transition to sustainability will therefore necessarily require non-technical (cultural, economic, political and social) ways of reducing the environmental footprint of the economy and bringing the demand for and supply of 'eco-capacity' more into balance. This footprint can be lightened in three ways: first, by altering the scale and profile of demand for final goods; secondly, by increasing available eco-capacity; thirdly, by increasing the efficiency with which eco-capacity is used.

Seeking to delineate sustainable circuits and spaces depends critically upon how 'sustainability' and 'sustainable development' are defined. These are contested concepts, with views as to how 'sustainability' ought to be defined ranging from pallid blue-green to dark deep green, emphasising that normative and political dimensions are never far from the agenda (McManus, 1996). For deep green ecologists, prioritising the preservation of nature is pre-eminent. Implementation of 'deep green' Gaian positions (Lovelock, 1988) would require significant reductions in material living standards and radical changes in the dominant social relations of production (Goodin, 1992). However, this could as well be associated with reactionary conservatism designed to preserve the *status quo* and protect the interests of the powerful as with progressive politics of radical social change (Pepper, 1984). Some dark green positions are unambiguously eco-fascist, with scant regard for the degraded living conditions and lifestyles of the majority of the world's population. They would condemn the vast majority of people to a miserable future, at best on the margins of physical existence. Ceasing to produce *all* toxins, hazardous wastes and radio-active materials would have disastrous consequences for public health and the well-being of millions of people (Harvey, 1996, 400). Such changes would be powerfully contested, as these would not, from the point of view of people, be sustainable spaces.

In contrast, paler blue-green perspectives envisage technological fixes within current relations of production, essentially trading off economic against environmental objectives, with the market as the prime resource allocation mechanism (Pearce et al., 1989). As such, they are incompatible with the creation of sustainable economic circuits and spaces. Between the deep green and blue-green perspectives are concepts of sustainability that accept the general legitimacy of markets as resource allocation mechanisms but recognise the need, in some circumstances, for non-market state regulation. In particular, such views acknowledge that 'some classes of natural resource and some environmental services are essential, non-substitutable and non-replicable. ... [As such] there must be ecological limits to economic activity' (Weaver et al., 2000, 33–4). This suggests that appropriate mixes of market and non-market regulation might create sustainable spaces.

The dominant views of 'sustainability' and sustainable development are grounded in a blue-green discourse of ecological modernisation, in claims that capital accumulation, profitable production and ecological sustainability are compatible goals (Hajer, 1995). As such, ecological modernisation theses conjoin doctrines of economic efficiency with those of ecological sustainability. The search for sustainable forms of development attempts to bridge the divide between technocratic/reductionist and ecocentric/holistic paradigms of development or eco-centric and anthropocentric views of the environment (O'Riordan, 1981). It reflects a desire to avoid a dangerous stand-off between advocates of economic growth and those of no growth, between optimistic cornucopians and pessimists for whom ecological disaster is immanent. Environmental quality is no longer seen as a luxurious positional good affordable only by the rich but is regarded as necessary for ecological survival and further economic development. Economic growth and environmental quality are now perceived as complementary objectives. Indeed, emphasising the importance of devising ecologically and environmentally sustainable forms of economic practice has become increasingly popular, often in seemingly unlikely quarters (for example, the World Bank, 1994).[3] However, evaluating competing claims as to what needs to be done to achieve sustainable development is complicated by the absence of consensus as to exactly what it means in practice.

Perhaps the most quoted – precisely because of its vagueness, a consequence of the political compromises that preceded it – definition of sustainable development is that of the United Nations World Commission on the Environment and Development (1987, 43): meeting 'the needs of the present, without compromising the ability of future generations to meet their own needs'. In broad terms, the Report seeks to work with the grain of, rather than radically challenge, the dominant logic of capitalist production. Thus defined, sustainability encompasses relations between the environment and economy and a commitment to equity, intra-generationally, inter-generationally and spatially. Development extends beyond quantitative growth in material outputs and incomes to include qualitative improvements in living and working conditions. Nevertheless, to remove the developmental gap between core and peripheral states of the capitalist economy, it advocated increasing manufacturing output five or ten-fold in the latter. While emphasising the need for qualitative changes in the character of growth, the Report was silent as to how this could be achieved. As such, it conspicuously failed to take account of the ecological consequences of such expansion in industrial output. The implications of this are considered more fully below.

10.6   Processes and policies for sustainability: innovation, eco-efficiency and eco-modernisation

In the past, there have been (neo)Malthusian fears, periodically expressed, about the apocalyptic consequences of finite stock resources, especially metallic minerals and fossil fuels, becoming exhausted, and imposing limits to growth and leading to economic decline (Meadows et al., 1972; Paley, 1952). While in the long run such resources will be exhausted, these fears have yet to be realised. This is primarily because successive technological fixes, allied to changing market conditions, have delayed resource exhaustion. Despite exhaustion of the richest ore deposits and the most accessible deposits of fossil fuels, regular improvements in 'exploitative technology' have allowed expanding production at declining prices. There are several dimensions to these changes.

First, significant improvements in methods of exploration and cartography have enabled the exploitation of previously unexplored territory. Remote sensing, based on satellite data, allowed identification of geological structures likely to contain significant mineral resources (Andrews, 1992). There have also been improvements in methods of ocean floor exploration (Hoagland, 1993).[4] On the other hand, as mineral ores become leaner, with a smaller ratio of the desired mineral(s) to waste material, or only available in less accessible locations, countervailing pressures increase the energy requirements and environmental consequences of raw material acquisition.

Secondly, mining and processing technologies have become more efficient. For example, more effective methods of extraction have extended the lives for existing oil and gas fields to 50 and 75 years, respectively. Processing technologies have also become more efficient: the energy inputs to smelt and refine metals have fallen markedly (Young, 1992). The glass and paper industries have significantly improved resource efficiencies through process innovations (Commission of the European Communities, 2001, 29). Furthermore, there are possibilities for further improvements. Process intensification in chemicals production and miniaturisation of production processes could increase materials and energy efficiency by 30% (Luseby, 1998).

Thirdly, recycling has expanded, especially in the spaces of advanced capitalism, with around 25% to 30% of materials previously fabricated into commodities or produced as waste by-products being recycled. Some companies and industries already achieve much higher levels. Steel produced by the electric arc route is produced entirely from scrap. The extent of recycling typically is a function of, *inter alia*, the relative prices of recycled and virgin materials so that the reasons for its extent 'appear to be not so much physical as economic' (Wernick et al., 1997, 146). As such, there is considerable scope for further recycling. Deconstruction involves disassembling buildings to salvage materials, de-manufacturing involves disassembling products into component parts, in both cases for re-use or recycling (Scharb, 2001, 18). Complex products such as automobiles could be designed to facilitate total recycling at the end of their useful lives. Such processes can have dramatic effects. For example, in the 1990s glass factories used as much as 90% recycled materials while 85% of cars produced by Honda and Toyota were recyclable. Often leading companies have pressed for more stringent environmental regulation as a source of competitive advantage (Day, 1998, 7). There is a very important exception in relation to potential for recycling, however – carbon fossil fuels. This may pose a major constraint on future patterns of economic activities. Moreover, recycling is not necessarily less energy intensive than production from virgin raw materials and may have other undesirable pollutant impacts (Hudson and Weaver, 1997).

Fourthly, a ubiquitous, or more widely available, material may be substituted for another that is less widely available. Aluminium can substitute for copper in electrical applications, although it is less conductive and so cannot be safely used in confined spaces in which there is a risk of overheating. Another option is to substitute an environmentally more benign material for a hazardous one. Ceramics based on widely available clays can substitute for metals (for example, in automobile engines) and sands for copper (as in the use of fibre optics in telecommunications). Such changes are often linked to process innovations and changes in the way in which products are designed and produced. The substitution of plastics for metals (for instance, in pipes or parts in automobiles) is more problematic, as plastics are typically produced from oil and may be difficult to recycle. In some circumstances, it may be possible to create synthetic substitutes or replace naturally occurring materials with new products that do not naturally occur as part of creating a 'second nature'.

Finally, finite stock resources may possibly be replaced with renewable flow resources. Coal, oil and gas could, in principle, be replaced as sources of primary energy generation and as raw materials in manufacturing processes by biomass, hydro-electric, solar, wave or wind power.[5] In particular, practically all the major commodity products of the synthetic organic chemicals sector could be produced, in principle, from plant materials. Methanol could become a potentially versatile starting compound in a sustainable economy, with important changes in its method of production (by new 'back-to-basics' primary processing technologies) and uses (producing hydrogen for fuel cells and as a starting compound for producing other chemicals). Because of this versatility, 'many analysts consider methanol to be a key chemical in the transition to sustainability'. While new primary conversion technologies are at an early stage of development, in principle they offer possibilities for 'converting bulk biomass and organic waste into versatile liquid or gaseous starting material from which whole families of final products could be produced' (Weaver et al., 2000, 238–45).

There is clearly scope for eco-restructuring, creating 'clean' production systems and dematerialising economic practices (Hudson and Weaver, 1997). For example, Germany, Japan and the USA reduced the material intensity of a unit of GDP by 20–30% over the last two decades of the twentieth century, although this was offset by aggregate growth in GDP. Consequently, total use of materials rose by almost 28% (Wernick et al., 1997, 139). The disjunction between per unit and aggregate trends emphasises the need to distinguish between process efficiency and product enhancement. Process efficiencies yield short-term benefits to firms, for example in waste reductions. Product enhancement emphasises product durability and service intensity. Day (1998, 5) argues that 'unfortunately' eco-efficiency conflates process efficiency and product enhancement, so that 'many firms are focusing on efficiency and calling it eco-efficiency, when it results in waste reduction alone'. However, such an approach 'is mistaken because it does not drive companies to innovate for a sustainable future. Innovation targeted towards the development of new products and new markets holds the greatest opportunity for business growth, and sustainable development as a business strategy can help firms drive this innovation'.

Companies pursuing sustainable development can move beyond the gains of process efficiencies and product enhancements to the construction of new markets for ecologically sustainable products in which they can grasp 'first mover advantage'. This emphasises Schumpeterian 'strong' competition as the key to sustainability via eco-modernisation. These companies have a vested interest in creating strong and rigorous frameworks of environmental regulation, within which they 'create visions of a sustainable future, anticipate latent or future consumer demands, and address them today'. Accordingly, there is a need for 'radical transformation' of technologies and economies (Day, 1998, 5), to maximise eco-benefit. In particular, 'three technology clusters in the areas of energy services, industrial materials supply and human nutrition ... account for much of total environmental stress' (Weaver et al., 2000, 43). Consequently, any transition to sustainability will need to address these three domains.

10.7 Moving onto sustainable technological trajectories

Weaver et al. (2000, 35) specify three necessary criteria for reducing the overall 'environmental footprint' of the global economy. First, depletion – there is no absolute exhaustion of resources. Secondly, pollution – there is no accumulation of pollutants or any lasting effects for future generations. Thirdly, encroachment – rates of loss must be no greater than rates of natural or anthropogenic restoration or replenishment.

However, making the globe a sustainable space also involves issues of distribution. The core territories of the global economy consume 25% of biomass, 80% of energy and 90% of metals (although there are major socio-spatial differences in the weight of environmental footprint within them). Put another way, if per capita car ownership in China, India and Indonesia rose to the global average of 90 cars/1,000 people, there would be effective demand for an additional 200 million cars. This sharply illustrates the tensions between the attractions for major fractions of capital of 'business as usual' and the ecological dangers that this encompasses if, as seems likely, such countries experience a mass consumption boom, as incomes rise and propensity to save declines. Remaining on the current developmental trajectory will lead to demand for eco-capacity exceeding supply by a factor of between 2 and 20 within 50 years. A transition to ecological 'sustainability' requires reducing per capita consumption of natural resources and emissions of pollutants by between 5% and 10% from the current levels in industrialised economies. Changes to established technologies will yield at most Factor 2 or 3 improvements in eco-efficiency and will be inadequate globally. Since sustainability involves greater equity, there is a strong moral imperative that the main responsibility for enhancing eco-efficiency should fall on those best able to do so. Thus eco-efficiency gains in the range Factor 10 to 50 will be required in the core territories of the capitalist economy. For example, oil consumption would need to be cut by a factor of 40, copper consumption by a factor of 30, and acid deposition by a factor of 50 (Weaver et al., 2000, 50). Changes of these magnitudes will require radical, path-breaking innovations – cultural, social and technological.

10.8 Sustainable economic geographies: practices, spaces and scales

10.8.1 *Sustainable production practices and spaces of production*

Adoption of, or indeed a return to, less materials and capital-intensive forms of production can reduce the intensity of the environmental footprint of agriculture, and increase productivity and product quality. Remarkable productivity gains can result from minimising chemical applications and maximising the use of natural regenerative processes, combined with local knowledges and skills (Pretty, 2001, reported in Bowring, 2003, 132). Such sustainable agricultural practices resulted in average crop yield increases of 50–100% for rain-fed crops, and 160% for small-to-medium root crops. Similarly, Cuba successfully converted from capital-intensive monoculture to low-input sustainable agriculture, a dramatic structural transformation necessitated by the post-1989 collapse of exports from central and eastern Europe that had previously met over half of Cuba's calorific requirements (Rosset and Benjamin, 1994).

Less intuitively obviously, agriculture could be made ecologically more efficient via food production 'factories' located in or near major urban areas. This would reduce transport and storage costs, enable a closer harmonisation of supply and demand and allow information about local needs, tastes and preferences to be better integrated into the production process. Consequently, only foods that met specified consumer requirements would be produced. Because production in controlled facilities would serve a small and localised community, feedback loops would allow just-in-time production, with ICT links between production units and retail outlets enabling supply and demand to be dovetailed in terms of variety, quality, quantity and timing. Creating precisely controlled localised growing environments would allow detailed specification of product characteristics. Such a system could yield major eco-efficiency gains, reducing fossil energy needs to almost zero, carbon dioxide inputs by a factor of 8 and water use by a factor of 18 (Weaver et al., 2000, 113–16).

Perhaps the greatest attention, however, has been focused upon sustainable forms and spaces of eco-industrial development (EID) and production. Based on a biological analogy, EID 'mimics the adaptive characteristics observed in nature by creating inter-firm relationships based on exchange and mutual gain'. As such, 'firms that successfully emulate nature's adaptive processes follow three important steps' (Ferri and Cefola, 2002, 34–8). First, they take a holistic view of their economic environment and identify potential network partners. Secondly, they find interdependencies and engage in resource exchanges. As well closing intra-firm materials loops via recycling, recovery or re-use of wastes, EID offers strategies to achieve greater eco-efficiency through 'economies of systems integration', in which 'partnerships between businesses meet common services, transportation and infrastructure needs'. Thirdly, they take advantage of exchanges to discover new products and process. Moreover, benefits spill over to local communities via environmental improvements, increased employment and more co-operative industrial relations. The emphasis is firmly upon EID creating win–win spaces.

EID takes one of two spatial forms, at one of two spatial scales: eco-industrial parks as discrete bounded (local) spaces or eco-industrial networks as discontinuous (trans-local) spaces. EID parks 'aspire to zero emission or closed loop manufacturing' and 'the total elimination of wastes' via exchanges of inputs and outputs (Spohn, 2002, 1). However, the emphasis 'on fostering networks among complementary firms and communities to optimise resource use and reduce economic and environmental costs' has increased recognition of the need 'to look at broader geographic ranges beyond a bounded industrial park to ensure economies of scale and sufficient supply of exchange materials' (Scharb, 2001, 1–2, 13). In practice, EID parks 'fall somewhere between the two extremes' of bounded local space and trans-local network (Ferri and Cefola, 2002, 35).

As such, EID can assume a variety of spatialities, from factory to firm, from eco-industrial park to regional (and beyond that national and international) networks, as companies enter into a variety of strategic alliances and linkages.

There is a number of examples of 'sustainable spaces', built around principles of EID. The eco-industrial park at Kalundborg, in Denmark, is typically seen as the pre-eminent example of successful eco-industrial development. Five industrial companies collaborate for mutual benefit, exchanging by-products. Although they have environmental benefits, these exchanges are based on bilateral commercial agreements 'built with economics in mind: the exchanges are not altruistic – they are driven by real profit incentives and the increased need for risk management' (Ferri and Cefola, 2002, 36). This has led to substantial cost saving which, along with improved environmental performance, confer a competitive advantage for participating companies.[6] In the USA and UK there is a number of initiatives that seeks to build on the experiences of Kalundborg (for example, see Cornell University, 2002; Scharb, 2001; Stone, 2002).

The most feasible locations for successful eco-industrial developments are big cities, which best meet four necessary conditions. First, there is an approximate balance between the demand for and supply of waste products. Secondly, inter-firm relationships based upon close individual connections or within an institutional framework that reduces transaction costs. Thirdly, sufficient compatibility between firms within close proximity to ensure stable quantities and qualities of by-products. Fourthly, regulations that encourage collaborative inter-firm relationships rather than the disposing of by-products as wastes.[7]

10.8.2   Sustainable consumption practices and spaces of consumption

Big cities also have potential to become spaces of more eco-efficient consumption. This can be exemplified with reference to cleaning and washing clothing and other household textiles. Currently, many more affluent households perform these tasks in the home, using automatic washing machines and tumble driers. Such appliances are major consumer durables for many households and a source of profits for companies that produce them. However, these activities have a heavy ecological footprint: they account for about 20% of household water and energy use in the Netherlands, for example (Weaver et al., 2000, 176). This partly reflects the way in which washing has taken on symbolic and ritualistic, as well as functional, dimensions. Alternative laundering techniques are inappropriate for use at the scale of the household. The potential eco-efficiency benefits in part depend upon more collective centralised provision. Assuming that cultural pressures for home-based systems could be overcome, the resultant scale economies would yield short- to medium-term eco-efficiency gains from recovering and re-using energy and materials and from matching cleaning treatment to need, reducing the resources used in the process per unit of laundry.

While there is scope for more eco-efficient urban production and distribution of food, creating more sustainable spaces will require greater and generalised gains in eco-efficiency in food production systems. Achieving these will necessitate more general societal changes in diet, food preferences and tastes. Consumption of meat contributes significantly to the environmental impacts of agriculture. Industrial meat production from grain-fed livestock is an ecologically inefficient and environmentally polluting method of producing edible energy and protein, leading to large, often concentrated, production of organic wastes and the loss of 80–90% of the contained nutritional value of the feedstock (Lappé, 1991). Added to this, the vast majority of transgenic crops are destined to become animal feedstock (Bowring, 2003). Furthermore, meat consumption is heavily skewed towards more affluent areas of the globe, in which it is both culturally sanctioned and affordable. For the two-thirds of the global population who have a predominantly vegetarian diet, meat is an unaffordable luxury, even if it is culturally sanctioned.

This example of potential dietary change highlights the complexities of shifting to more sustainable economic practices. Novel foods could enable protein to be produced with Factor 20 to 30 improvements in resource productivity and at substantially lower economic cost. However, there are major cultural, economic and social barriers to their adoption. Meeting nutritional needs is not the only function of foods, which provide satisfaction through their aromas, flavours and textures, and help shape bodies and identities. Moreover, 'they also say something about us. Foods are used both to confer and confirm social standing. Important relationships and family occasions are marked by eating important foods.' As a result, 'all in all, the concerns of consumers over conventional foods in eating norms and habits constitute significant barriers to dietary change' (Weaver et al., 2000, 121–2). Furthermore, these have been reinforced by the incidence of specific disease transmission from animals to humans through the food chain, creating concerns about human health, in addition to those about the ethical implications of food choices in terms of animal welfare in industrialised farming systems and the 'hidden' relations of exploitative human labour in commodity food chains (Jackson, 2002). Consequently, pressures against innovative and potentially eco-efficient food products have been amplified, although the shift to vegetarianism does create such gains as a by-product of not eating meat.

10.8.3   *Sustainable flows and spaces*

Constructing sustainable economic spaces requires radical shifts in transport technologies and, over the longer term, land use patterns. Sustained increases in areas of urban land, land converted to road space and per capita car ownership are indicative of the intimate link between automobility (Urry, 1999) and lifestyle for many people, for whom mobility is an important element of their quality of life. In turn, this reflects the power of the 'road lobby' (Hamer, 1974) to promote its interests around the manufacture of cars and the construction of roads and other infrastructure on which to drive them. Land use patterns and transport demand and supply arrangements have co-evolved so that the capacity to be highly mobile and the demand for mobility have been mutually supportive. The resultant aggregate environmental impacts of transport require reduction as part of any transition to sustainability. So too do the inequalities in mobility that have resulted from prioritising the private car as a mode of transport over much of the world.

There are, in principle, four ways of reducing the environmental impacts of transport. First, by demand management, reducing aggregate flows. Secondly, by increasing vehicle loads and optimising routes. Thirdly, by changing modal splits. The scale of car use and resultant inequalities in mobility and massive carbon dioxide discharges into the atmosphere provide compelling reasons for such a switch. However, this will require radical changes to the mobilities and spatialities of everyday life that have become socially dominant and which many people both expect and/or are required to perform in economy and society. Fourthly, by changing transport technologies. This implies a radical shift from the vehicle-mounted internal combustion engine as a source of on-board power because it 'contributes most to the overall inefficiency of the well-to-wheel chain'. Moreover, existing designs for the internal combustion engine are 'close to the theoretical efficiency limits. To increase the efficiency of the transport source-to-service chain as a whole, there is a need to develop an alternative technology'. While it is 'too early to discount any [possible] options … hydrogen or hydrogen-rich fuel cell technologies offer the greatest long-term potential to compete effectively on cost, performance, efficiency and sustainability criteria' (Weaver et al., 2000, 251–67). Such technologies could be commercially viable within 15 years.

For many people, constructing cities and regions as sustainable spaces of consumption, movement and production will require radically changing lifestyles as journeys to work, to shop, and for purposes of recreation are reshaped. For this to be possible, any meaningful longer-term transition to 'sustainability' will require major changes to the spatial arrangement of built environments, the relative locations of spaces of work, exchange, leisure and residences, and commensurate changes in peoples' activity patterns and spaces. In brief, it will require a shift from built environments designed to maximise the movements required to go to work, shop and play to environments designed to minimise such movements. Planning and designing such built environments will drastically alter the relative locations of spaces of dwelling, work and so on and also the scales at which these activities occur. However, considerable inertia is built into built environments precisely because they are constituted via major outlays of fixed capital (both private and public sector investment), amortised or depreciated over long periods of time, typically decades. Short of the drastic option of mass destruction due to warfare or catastrophic environmental changes, there are powerful economic imperatives to preserve the socio-spatial structures of cities and regions, or at least slow the pace of change so that it does not endanger existing fixed capital investments and steer it so that it provides further scope for capital accumulation.

Recognising these barriers arising from the dominant social relations of the economy, the potential for such changes to mobility and movement and to the geometry and scale of built environments is, to a degree, already present. For example, developments in ICTs, especially the Internet, 'offer the potential for large-scale vehicle commuting to be replaced by virtual offices in the home, at "village" sites or on-board vehicles connected by dial-up communications' (Button and Taylor, 2001, 30). Linked to this, the growth of e-commerce and the reduced need physically to visit retail outlets is seen as a mechanism for reducing non-work travel. Whether this potential really is 'large scale' and, if it is, whether it will be realised remains a contingent matter, however. For example, Button and Taylor cite surveys of commuters in California, which reveal that only 2% wanted a zero to two minute commute while almost 50% preferred a commute of 30 minutes or more, suggesting resistance to the erosion of auto-mobile-based lifestyles and the spatial separation of spaces of work and residence. There are, however, potential counter-tendencies, with unwanted effects. For example, the growth of mass customisation and small-batch production, which has also been facilitated by developments in ICTs, 'have inevitable implications for delivery patterns, involving smaller consignments and more frequent deliveries ... [which may increase] the efficiency of the freight transportation business by as much as 15–20%. This saving will itself keep the costs of transportation down although the social and environmental implications may be somewhat different' (Button and Taylor, 2001, 33). This further indicates the complexity of seeking to move on to more sustainable developmental trajectories and create sustainable spaces.

10.8.4 *Regulation and the creation of sustainable flows and spaces: steering transitions to eco-capitalism or eco-socialism?*

The way in which sustainable spaces are created clearly depends upon the underlying concept of sustainability. In general, 'weak' sustainability is the dominant concept of sustainability informing public policy (Springett, 2003). As a result, the dominant conception is of spaces of (very) weak sustainability, underpinned by a view of 'business as usual', as eco-modernisation regards producing profits and producing sustainable economies as compatible. While there may be improvements in eco-efficiency, this is firmly within an eco-modernisation paradigm that leaves the basic social relations of the economy untouched. The assumptions underlying such beliefs are neatly summarised by the Commission for the European Communities (2001, 4): 'if policy makers create the right conditions and encourage citizens and businesses to integrate environmental and social considerations in all their activities, policies for sustainable development will create many "win-win" situations, good for the economy, employment and the environment'.

Regulatory change can, within limits, encourage changes in product design, product innovation and consumer behaviour and lead to significant improvement in eco-efficiency. For example, energy-efficient condensing boilers, invented in the 1970s in the Netherlands, are virtually absent in UK dwellings. By 2007, however, new environmental regulations will require all new and replacement domestic boilers in the UK to be condensing boilers. This will create a substantial (around 1.3 million per annum) market for these 'green' commodities. Again an European Union directive issued in 2002 requires that by 2006 manufacturers of electrical equipment take back and recycle at least 50% of old scrapped equipment (such as refrigerators or PCs). By this date national governments in EU states will be required to collect an average of 4 kg of 'waste' electrical goods per household in order 'to deter consumers from discarding old computers, toasters and other electrical appliances with ordinary household refuse'. Moreover, 'individual manufacturers will be responsible for organising the disposal, re-cycling and re-use of the goods they put on the market after September 2005, creating an incentive for "greener" design' (Houlder, 2002).

The dominant eco-modernist view of 'sustainability' and the resultant regulatory frameworks that flow from it envisages that eco-capitalism *is* possible. Technological innovation, suitably steered by regulation and market construction, will lead to a new sustainable trajectory of development. In this context, it is important to enquire about the limits to technologies (Weaver et al., 2000, 50). First, there are limits set by chemical and physical laws that specify absolute maximum potentials for qualities such as conversion efficiencies. For example, there is an absolute minimum amount of carbon needed to reduce iron ore to pig iron. Secondly, there are 'configuration and context-dependent technological limits, which in practice are more pervasive and constraining'. Consequently, the major limits are institutional and social. As such, progress towards radical improvement in eco-efficiency 'will depend not only on meeting the technological challenges, but also on co-evolutionary developments in policies, markets, attitudes and behaviour. Research and development and innovation efforts will have to be directed to all these challenges.'

Thus, shifting towards sustainability requires interrelated changes to economic and market structures, production and consumption profiles, technologies, institutions and organisational arrangements, especially given the long lead times – maybe 50 years – needed for innovations that involve systemic and paradigmatic changes. For example, the replacement of water-based with chemical solvent-based technologies for cleaning textiles will depend upon market construction and regulation. This is because the demand to develop the relevant technologies 'can only come from large service providers who, if they have a sufficient client base, will be driven in this direction by competitive pressures to cut costs'. Within the limits of a capitalist economy, this will require large-scale provision at centralised facilities. In turn, longer-term development of such alternative technologies depends upon establishing short- to medium-term demand for professional laundry services 'to drive the R&D process' (Weaver et al., 2000, 202).

The long lead times needed for securing qualitative systemic change requires that innovations of different types and over different time horizons are compatible and consistent with long-term objectives. For convenience, these may be categorised as short-, medium- and long-term. Short-term innovations (over a time horizon of five or fewer years) are mostly concerned with 'good housekeeping' of existing technologies – for example, 'end-of-pipe' additions. Medium-term innovations (5–20 years) relate to process and product integrated technological developments. Long-term innovations (50 or more years) seek radical change to technological and organisation arrangements. Whereas the first two types require single-loop learning, the third requires double-loop learning as a necessary condition for the possibility of radical change.

In short, for Weaver et al., as for many others, eco-modernisation offers a feasible route to sustainable trajectories and spaces within the social relations of a capitalist economy. The Netherlands Sustainable Technology Development Programme in the 1990s suggests that sustainable technologies are 'not easily developed in all fields of need and substantial, conscious and consistent efforts are needed to search for sustainable technologies', but such technologies *can* nevertheless be developed (Weaver et al., 2000, 286). Similarly, Cornell University (2002, 4) argues that 'an eco-system view that looks for profitable niches in the current production web can absorb sinkholes that currently exist in the local eco-system and thus improve overall eco-system health'. More generally, production of profits and production of sustainability *are* seen as mutually compatible and attainable goals. Eco-modernisation suggests that the specific tensions between economy and environment and the more general underlying contradictions of capitalist development can be contained by appropriate technological, institutional, regulatory and behavioural changes.

Others, however, dispute that such a happy 'win–win' state is feasible precisely because of the inherent structural contradictions of the social relations of capital. Seeking to address the ecological footprint of economic practices and the distribution of economic and environmental costs and benefits so that the former is ecologically tolerable and the latter meet criteria of ethics and environmental and social justice (Harvey, 1996) is deeply problematic. It remains an open question as to whether sustainable capitalism is possible. For J. O'Connor (1994, 154–5) 'the short answer to the question is "no", while the longer answer is "probably not". For those who dispute that eco-capitalism is a feasible project, the awkward question of what an eco-socialist alternative would look like is difficult to avoid – and difficult satisfactorily to answer. While the prospects for a 'sustainable capitalism' are very remote, those of 'some kind of "ecological socialism"' are not much better. While not particularly helpful, this does have the merit of avoiding the pretence that there are *any* easy answers. Even so, there may be some scope for the constructing 'local' sustainable spaces grounded in alternatives to the social relationship of the mainstream capitalist economy. For example, local currency systems and Local Exchange Trading Systems seek to create more localised systems of production, exchange and consumption, based around non-capitalist concepts of value and local currencies that have only a restricted spatial validity.

Attempts to develop localised economies around non-capitalist social relations often centre on projects to revalorise and recycle commodities discarded by others because they have reached the end of their original socially useful or 'fashionable' life. Others focus on projects that seek to enhance environmental quality and revalorise the built and natural environments. The plethora of markets (from old-style 'flea markets' to more recent innovations such as car boot sales) in which things are recycled to new users suggest that, if only at the margins, there is scope for more sustainable consumption practices. Such alternatives are admirable in themselves (not least in demonstrating that there *are* alternative forms of practice) but they remain local experiments on the fringes, or in the interstices, of the mainstream capitalist economy rather than systemic alternatives to it. Indeed, it may be that their existence helps legitimate the capitalist mainstream, evidence that it is tolerant of diversity and difference, rather than their being the first faltering steps towards an alternative – maybe eco-socialist – mainstream.

10.9   Summary and conclusions

The capitalist economic system lacks 'natural' regulatory mechanisms and effective social regulatory mechanisms to control materials dispersion. If anything state socialism imposed even less constraints. Economic activity is essentially and unavoidably dissipative. Consequently, the processes of production, exchange, consumption and transportation that constitute the economy unavoidably have 'unintended consequences' and create unwanted effects in the form of waste products. This conclusion holds irrespective of the social relations of the economy, although the extent and form of these unintended outcomes and the production of wastes varies within and between different social forms of economic organisation. This results in a complex mosaic of spaces of wastes, in complex spatialities of wastes as some effects are intensely localised while others become one of the few genuine examples of processes of globalisation. Industrial production and consumption freely dissipate a wide array of chemicals, some non-existent in nature, and many of which exceed natural flows by several orders of magnitude (Jackson, 1995, 7–20).

Depletion and dissipation of carbon-based fuels and other minerals is unavoidable. Moving to a de-carbonised economy may conceivably be possible within capitalist relations of production, provided that the activities associated with it satisfy the normal profitability criteria and fall within socially and politically acceptable limits. Then again, it may not be. Thus the extent to which systems of social and political regulation and governance are constructed to ensure that economies move to more environmentally – and socially – sustainable trajectories is uncertain. While there are strong reasons for believing that some form of eco-capitalism is improbable, the chances of a systemic non-capitalist alternative are at least equally distant.

Finally, irrespective of the social relations of the economy, there are good reasons for exercising the precautionary principle in seeking to create sustainable spaces. Creating a sustainable local space is clearly problematic. Creating a sustainable global space poses even greater challenges. The global space can be conceptualised as a mosaic of complex economies, an open system that maintains an orderly state by capturing and using radiant solar energy. Because the system is complex and non-linear, its dynamic behaviour is potentially chaotic. System stability, which depends upon complex feedback loops, 'is only assured when the system-state remains within a certain range. The resilience of the system – its tendency to remain within its original domain – is indeterminate' (Weaver et al., 2000, 33). In short, given both our partial knowledge of the complexities of economy/environmental systems interactions, and the emergent (and by definition currently unknown) properties of such complex co-determining systems, there are strong grounds for proceeding cautiously. For the 'indeterminacy of the time path of a complex system … implies the impossibility of predicting in advance, and perhaps in any way controlling, the "inputs" that this system will provide to others co-dependent upon it. Autonomy and openness are inherently linked to indeterminacy and incomplete control' (M. O'Connor, 1994a, 66). Since administrative rationality cannot cope with truly complex problems (Dryzek, 1994, 181), there are profound implications for policy choices and modes of policy implementation in seeking to create sustainable spaces, at all spatial scales.

Notes

1 The environmental legacies of state socialism are equally serious. For example, the Kula Peninsula of Russia is heavily polluted by wastes from 250 nuclear power stations and from decommissioning nuclear submarines (Tatko and Robinson, 2002).

2 Despite claims that dismantling ships would create jobs, memories of a history of polluting industry have led to strong local opposition.

3 Although for some the term 'sustainable development' is an oxymoron (Goldsmith, 1992).

4 Establishing international legal frameworks to enable exploitation of sea-bed resources, such as manganese nodules, raises complex issues, however.

5 However, only 0.02% of global energy supply is generated from water or wind.
6 Major petro-chemical complexes are characterised by such product interchanges (Hudson, 1983).
7 Recognition of such issues is not new, however. In 1920 Talbot (cited in Scharb, 2001, 22) identified the key issue as follows: 'waste must be forthcoming in a steady stream of uniform volume to justify its exploitation, and the fashioning of these streams is the supreme difficulty'.

References

Aglietta M, 1979, *A Theory of Capitalist Regulation: The US Experience*, New Left Books, London.

Aglietta M, 1999, 'Capitalism at the turn of the century: regulation theory and the challenge of social change', *New Left Review*, 232, 41–96.

Allen J, 2002, 'Symbolic economies: the "culturalization" of economic knowledge', in Du Gay P and Pryke M (eds), *Cultural Economy*, Sage, London, 39–58.

Allen J, Cochrane A and Massey D, 1998, *Re-thinking the Region*, Routledge, London.

Allen J and Henry N, 1997, 'Ulrich Beck's Risk Society at work: labour and employment in the contract service industries', *Transactions of the Institute of British Geographers*, NS, 22, 180–96.

Altvater E, 1990, 'Fordist and post-Fordist international division of labour and monetary regimes', Paper presented to the Conference on Pathways to Industrialization and Regional Development in the 1990's, University of California at Los Angeles, 14–18 March.

Altvater E, 1994, 'Ecological and economic modalities of time and space', in O'Connor M (ed.), *Is Capitalism Sustainable?*, Guilford, New York, 77–90.

Amin A, 1998, 'An institutionalist perspective on regional economic development', Paper presented to the RGS Economic Geography Research Group Seminar, Institutions and Governance, 3 July, University College London.

Amin A, 2000, 'Organisational learning through communities of practice', Paper presented to the Workshop on The Firm in Economic Geography, University of Portsmouth, 9–11 March.

Amin A and Cohendet P, 1997, 'Learning and adaptation in decentralised business networks', Paper presented to the European Management Organisations in Transition Final Conference, Stresa, 11–13 September.

Amin A and Thrift N, 1992, 'Neo-Marshallian nodes in global governance', *International Journal of Urban and Regional Research*, 16: 571–87.

Amin A and Thrift N, 2004, 'Cultural economy: the genealogy of an idea', in Amin A and Thrift N (eds), *Cultural Economy: A Reader*, Sage, London (forthcoming).

Amin S, 1977, *Unequal Development*, Harvester, Brighton.

Anderson B, 1982, *Imagined Communities*, Verso, London.

Anderson J, 1995, 'The exaggerated death of the nation state', in Anderson J, Brook C and Cochrane A (eds), *A Global World*, Oxford University Press, Oxford, 65–112.

Anderson P, 1984, *In the Tracks of Historical Materialism*, Verso, London.

Andrews C B, 1992, 'Mineral sector technologies: policy implications for developing countries', *Natural Resources Forum*, 16, 212–20.

Angel D P, 1989, 'The labour market for engineers in the US semi-conductor industry', *Economic Geography*, 65, 99–112.

Aoyama Y, 2001, 'The information society, Japanese style: corner stores as hubs for e-commerce access', in Leinbach T R and Brunn S D (eds), *Worlds of E-Commerce*, Wiley, Chichester, 109–28.

Appadurai A (ed.), 1986, *The Social Life of Things*, Cambridge University Press, Cambridge.

Arlacchi P, 1983, *Mafia, Peasants and Great Estates*, Cambridge University Press, Cambridge.

Arora A and Gamborella A, 1994, 'The changing technology of technological change: general and abstract knowledge and the division of innovative labour', *Research Policy*, 23, 523–32.

Asheim B, 2000, 'Industrial districts: the contributions of Marshall and beyond', in Clark G L, Feldman M P and Gertler M (eds), *The Oxford Handbook of Economic Geography*, Oxford University Press, Oxford, 413–31.

Athreye S, 1998, 'On markets in knowledge', *ESRC Centre for Business Research, Working Paper 83*, University of Cambridge, Cambridge.

August O, 2002, 'Workers abused in sweatshops of China's economic miracle', *The Times*, 24 December.

Baker A, 2000, *Serious Shopping*, Free Association Books, London.

BAN (Basle Action Network), 2003, *Needless Risk: The Bush Administration's Scheme to Export Toxic Waste Ships to Europe*, BAN, Basle, 63 pp.

Barthes P, 1985, *The Fashion System*, Jonathan Cape, London.

Baudrillard J, 1988, 'Consumer society', in Poster M (ed.), *Selected Writings*, Polity Press, Cambridge.

Bauman Z, 1992, *Intimations of Postmodernity, Routledge*, London.

Bauman Z, 2001, *Community: Seeking Safety in an Insecure World*, Polity Press, Cambridge.

Beck U, 1992, *Risk Society: Towards a New Modernity*, Sage, London.

Benton T, 1989, 'Marxism and natural limits: an ecological critique and reconstruction', *New Left Review*, 178, 51–86.

Berman B and Evans J R, 1998, *Retail Management: A Strategic Approach* (7th edition), Prentice-Hall, Englewood Cliffs, N J.

Berry B J L, 1967, *Geography of Market Centres and Retail Distribution*, Prentice-Hall, Englewood Cliffs, NJ.

Best M H, 1990, *The New Competition: Institutions of Industrial Restructuring*, Polity Press, Cambridge.

Beynon H, 1973, *Working for Ford*, Penguin, Harmondsworth.

Beynon H, Cox A and Hudson R, 2000, *Digging up Trouble: The Environment, Protest and Opencast Coal Mining*, Rivers Oram, London.

Beynon H, Hudson R and Sadler D, 1991, *A Tale of Two Industries: The Contraction of Coal and Steel in North East England*, Open University Press, Milton Keynes.

Beynon H, Hudson R and Sadler D, 1994, *A Place Called Teesside: A Locality in a Global Economy*, Edinburgh University Press, Edinburgh.

Blomley N, 1996, '"I'd like to dress her all over": masculinity, power and retail capital', in Wrigley N and Lowe M (eds), *Retailing, Consumption and Capital: Towards a New Retail Geography*, Addison Wesley Longman, Harlow, 238–56.

Blowers A, 1984, *Something in the Air: Corporate Power and the Environment*, Harper & Row, London.

Bourdieu P, 1977, *Outline of a Theory of Practice*, Cambridge University Press, Cambridge.

Bourdieu P, 1981, 'Men and machines', in Knorr-Cetina K and Cicourcel L (eds), *Advances in Social Theory and Methodology*, Routledge and Kegan Paul, Boston, MA, 304–18.

Bourdieu P, 1984, *Distinction: A Social Critique of the Judgement of Taste*, Routledge and Kegan Paul, London.

Bourdieu P, 1987, 'Marginalia', in Schrift A (ed.), *The Logic of the Gift*, Routledge, London, 231–44.

Bowring F, 2003, 'Manufacturing scarcity: food biotechnology and the life-sciences industry', *Capital and Class*, 79, 107–44.

Braczisch H-J, Cooke P and Heindenreich M (eds), 1998, *Regional Innovation Systems*, UCL Press, London.

Brenner N, Jessop B, Jones M and MacLeod G (eds), 2003, *State/Space: A Reader*, Blackwell, Oxford.

Brunetta R and Ceci A, 1998, 'Underground employment in Italy: its causes, its extent and the costs and benefits of regularisation', *Review of Economic Conditions in Italy*, 2, 257–90.

Brun-Rovet M, 2002, 'Children clean up to the tune of £1,200 a year', *Financial Times*, 10 May.

Budd L and Whimster S, 1992, *Global Finance and Urban Living: The Case of London*, Routledge, London.

Bullard R, 1994, *Unequal Protection: Environmental Justice and Communities of Color*, Sierra Club Books, San Francisco, CA.

Burkett P, 1997, 'Nature's 'free gifts' and the ecological significance of value', *Capital and Class*, 68, 89–110.

Button K and Taylor S, 2001, 'Towards an economics of the Internet and e-commerce', in Leinbach T R and Brunn S D (eds), *Worlds of E-Commerce*, Wiley, Chichester, 27–44.

CAFOD, 2004, *Clean up your Computer: Working Conditions in the Electronics Sector*, CAFOD, London.

Calton W R, Jr, 1989, 'Cargoism and technology and the relationship of these concepts to important issues such as toxic waste disposal sites', in Peck D (ed.), *Psychosocial Effects of Hazardous Toxic Waste Disposal on Communities*, Charles C Thomas, Springfield, IL, 99–117.

Castells M, 1996, *The Rise of the Network Society*, Blackwell, Oxford.

Castells M and Portes R, 1989, 'World underneath: the origins, dynamics and affects of the informal economy', in Portes A, Castells M and Benton L A (eds),

The Informal Economy: Studies in Advanced and Less Developed Countries, Johns Hopkins University Press, Baltimore, MD, 1–30.

Castree N, 1999, 'Envisioning capitalism: geography and the renewal of Marxism political economy', *Transactions of the Institute of British Geographers*, NS, 24, 137–58.

Cerny P, 1990, *The Changing Architecture of Politics: Structure, Ageing and the Future of the State*, Sage, London.

Champion A G (ed.), 1989, *Counterurbanization: The Changing Pace and Nature of Population Deconcentration*, Arnold, London.

Chaney D, 1983, 'The department store as a cultural form', *Theory, Culture and Society*, 1, 22–31.

Chaney D, 1990, 'Subtopia in Gateshead: the Metrocentre as a cultural form', *Theory, Culture and Society*, 7, 49–68.

Chesnais F, 1993, 'Globalisation, world oligopoly and some of their implications', in Humbert M (ed.), *The Impact of Globalisation on Europe's Firms and Regions*, Pinter, London, 12–21.

Christopherson S, 1996, 'The production of consumption: retail restructuring and labour demand in the USA', in Wrigley N and Lowe M (eds), *Retailing, Consumption and Capital: Towards a New Retail Geography*, Addison Wesley Longman, Harlow, 159–77.

Clancy M, 1998, 'Commodity chains, services and development: theory and preliminary evidence from the tourism industry', *Review of International Political Economy*, 5, 122–48.

Clark G L, 1992, 'Real regulation: the administrative state', *Environment and Planning A*, 24, 615–27.

Clarke A J, 1997a, 'Window shopping at home: classified, catalogues and new consumer skills', in Miller D (ed.), *Material Cultures*, UCL Press, London, 73–99.

Clarke A J, 1997b, 'Tupperware: suburbia, sociality and mass consumption', in Silverstone R (ed.), *Visions of Suburbia*, Routledge, London, 132–60.

Coe N, 1997, 'US transnationals and the Irish software industry: assessing the nature, quality and stability of a new wave of foreign direct investment', *European Urban and Regional Studies*, 4, 211–30.

Cole G, 1999, 'Engineers welcome the age of knowledge', *Financial Times*, 2 June.

Commission of the European Communities, 2001, *Consultation Paper for the Preparation of a European Union Strategy for Sustainable Development*, SEC (2001) 517, Brussels, 57 pp.

Commission on the Macroeconomics of Health, 2001, *Macroeconomics and Health: Investing in Health for Economic Development*, World Health Organisation, Geneva.

Cooke, P and Wells P, 1991, 'Uneasy alliances: the spatial development of computing and communication markets', *Regional Studies*, 25, 345–54.

Cornell University, 2002, 'Cornell's perspective on eco-industrial parks', available at http://www.cfe.cornell.edu/wei/cupersp.html [accessed 3 December 2002].

Correa C, 1995, 'Sovereign and property rights over plant genetic resources', *Agriculture and Human Values*, 12, 58–79.

Crang P, 1994, 'It's showtime: on the workplace geographies of display in a restaurant in south east England', *Society and Space*, 12, 675–704.

Crang P, 1996, 'Displacement, consumption and identity', *Environment and Planning A*, 28, 47–67.

Crang P, 1997, 'Introduction: cultural turns and the (re)constitution of economic geography' in Lee R and Wills J (eds), *Geographies of Economies*, Arnold, London, 3–15.

Crawford M, 1992, 'The world in a shopping mall', in Sorkin M (ed.), *Variations on a Theme Park: the New American City and the End of Public Space*, Noonday Press, New York, 3–30.

Crewe L, 1996, 'Material culture: embedded firms, organizational networks and the local economic development of a fashion quarter', *Regional Studies*, 30, 257–72.

Crewe L, 2000, 'Geographies of retailing and consumption', *Progress in Human Geography*, 24, 275–90.

Crewe L, 2001, 'The besieged body: geographies of retailing and consumption', *Progress in Human Geography*, 25, 629–40.

Crewe L and Davenport E, 1992, 'The puppet show: changing buyer–supplier relations within clothing retailing', *Transactions of the Institute of British Geographers*, 17, 183–97.

Crewe L and Gregson N, 1998, 'Tales of the unexpected: exploring car boot sales as marginal spaces of contemporary consumption', *Transactions of the Institute of British Geographers*, 23, 39–53.

Crewe L and Lowe M, 1995, 'Gap on the map? Towards a geography of consumption and identity', *Environment and Planning A*, 27, 1877–98.

Crewe L and Lowe M, 1996, 'United colours? Globalisation and localisation tendencies in fashion retailing', in Wrigley N and Lowe M (eds), *Retailing, Consumption and Capital: Towards a New Retail Geography*, Addison Wesley Longman, Harlow, 271–83.

Damette F, 1980, 'The regional framework of monopoly exploitation', in Carney J, Hudson R and Lewis J (eds), *Regions in Crisis: New Perspectives in European Regional Theory*, Croom Helm, London, 76–92.

Davies G, 1994, *A History of Money from Ancient Times to the Present Day*, University of Wales Press, Cardiff.

Day R M, 1998, 'Beyond eco-efficiency: sustainability as a driver for innovation', World Resources Institute, 9 pp. Available at http://www.wri.org/meb/sei/beyond.html [accessed 3 December 2002].

Dean M, 1999, *Governmentality*, Sage, London.

Delapierre M and Zimmerman J B, 1993, 'From scale to network effects in the computer industry: implications for an industrial policy', in Humbert M (ed.), *The Impact of Globalisation on Europe's Firms and Industries*, Pinter, London, 76–83.

Deléage J P, 1994, 'Eco-Marxist critique of political economy', in O'Connor M (ed.), *Is Capitalism Sustainable?*, Guilford, New York, 37–52.

Dicken P, 1998, *Global Shift* (3rd edition), Sage, London.

Dicken P, Forsgren M and Malmberg A, 1994, 'The local embeddedness of transnational corporations', in Amin A and Thrift N (eds), *Globalization,*

Institutions and Regional Development in Europe, Oxford University Press, Oxford, 23–45.

Dicken P and Thrift N, 1992, 'The organization of production and the production of organization: why business enterprises matter in the study of geographical industrialization', *Transactions of the Institute of British Geographers*, NS, 17, 270–91.

Doane M, 1987, *The Desire to Desire*, Indiana University Press, Bloomington, IN.

Dodd N, 1994, *The Sociology of Money: Economics, Reason and Contemporary Society*, Polity Press, Cambridge.

Domosh M, 1996, 'The feminized retail landscape: gender ideology and consumer culture in nineteenth-century New York City', in Wrigley N and Lowe M (eds), *Retailing, Consumption and Capital: Towards a New Retail Geography*, Addison Wesley Longman, Harlow, 257–70.

Donzelot J, 1991, 'Pleasure in work', in Burchell G, Gordon C and Miller P (eds), *The Foucault Effect: Studies in Governmentality*, Harvester Wheatsheaf, Hemel Hempstead.

Douglass D and Krieger J, 1983, *A Miner's Life*, Routledge and Kegan Paul, London.

Dowling R, 1993, 'Femininity, place and commodities: a retail case study', *Antipode*, 25, 295–319.

Dryzek J S, 1994, 'Ecology as discursive democracy: beyond liberal capitalism and the administrative state', in O'Connor M (ed.), *Is Capitalism Sustainable?*, Guilford, New York, 176–97.

Du Gay P and Pryke M, 2002, 'Cultural economy: an introduction', in Du Gay P and Pryke M (eds), *Cultural Economy*, Sage, London, 1–20.

Dugger W M, 2000, 'Deception and inequality: the enabling myth concept', in Pullin R (ed.), *Capitalism, Socialism and Radical Political Economy*, Edward Elgar, Cheltenham, 66–80.

Dunford M, 1990, 'Theories of regulation', *Society and Space*, 8, 297–321.

Dunford M, 2004, 'Comparative economic performance, inequality and the market-led re-making of Europe', *European Urban and Regional Studies* (forthcoming).

Economic Policy Institute, 2000, *Manufacturing Advantage: Why High Performance Work Systems Pay Off*, Cornell University Press, Ithaca, NY.

Edgecliff-Johnson A, 1998, 'Crosfield sale points to more ICI disposals', *Financial Times*, 3 April.

Elson D, 1979, 'The value theory of labour', in Elson D (ed.), *Value: The Representation of Labour*, CSE Books, London, 115–80.

Ewen S, 1976, *Captains of Consciousness: Advertising, the Social Roots of the Consumer Culture*, McGraw-Hill, New York.

Ewen S, 1988, *All Consuming Images*, Basic Books, London.

Featherstone M, 1991, *Consumer Culture and Postmodernism*, Sage, London.

Featherstone M, 1998, 'The *flâneur*, the city and virtual public life', *Urban Studies*, 35, 909–25.

Fernie J, 1998, 'The breaking of the fourth wave: recent out-of-town retail developments in Britain', *International Journal of Retail, Distribution and Consumer Research*, 8, 303–17.

Fernie J, Moore C M, Lawrie A and Hallsworth A, 1997, 'The internationalization of the high fashion brand: the case of Central London', *Journal of Product and Brand Management*, 6, 151–62.

Ferri M R and Cefola J, 2002, 'A case for eco-industrial development', available at http://www.cfe.cornell.edu/wei/ [accessed 3 December 2002].

Fine B and Leopold E, 1993, *The World of Consumption*, Routledge, London.

Fitzsimmons M, Glaser J, Montemor R, Pincett S and Rajan S C, 1994, 'Environmentalism as the liberal state', in O'Connor M (ed.), *Is Capitalism Sustainable?*, Guilford, New York, 198–216.

Flamm K, 1993, 'Coping with strategic competition in semiconductors: the EC model as an international framework', in Humbert M (ed.), *The Impact of Globalisation on Europe's Firms and Regions*, Pinter, London, 64–75.

Florida R, 1995, 'The industrial transformation of the Great Lakes region', in Cooke P (ed.), *The Rise of the Rustbelt*, University of London Press, London, 162–76.

Florida R, 2002, *The Rise of the Creative Class and How It's Transforming Work, Leisure, Community and Everyday Life*, Basic Books, New York.

Foremski T, London S and Waters S, 2003, 'Smaller chips, bigger stakes: how new, monster plants will transform the industry', *Financial Times*, 11 June.

Franklin S, Lury C and Stacey J, 2000, *Global Nature, Global Culture*, Sage, London.

Friedman A, 1977, *Industry and Labour*, Macmillan, London.

Frosch R A, 1997, 'Towards the end of waste: reflections on a new ecology of industry', in Ausubel J H and Langford H D (eds), *Technological Trajectories and the Human Environment*, National Academy Press, Washington, DC, 157–67.

Fucini J and Fucini S, 1990, *Working for the Japanese: Inside Mazda's American Auto Plant*, Free Press, Toronto.

Fyfe N, 1998, 'Introduction', in Fyfe N (ed.), *Images of the Street*, Routledge, London, 1–18.

Garofoli G, 2002, 'Local development in Europe: theoretical models and international comparisons', *European Urban and Regional Studies*, 9, 225–40.

Garonna P, 1998, 'The crisis of the employment system in Italy', *Review of Economic Conditions in Italy*, 2, 219–56.

Garreau J, 1991, *Edge City: Life on the New Frontier*, Anchor Books, New York.

George S, 1992, *The Debt Boomerang: How Third World Debt Harms Us All*, Pluto Press, London.

Georgescu-Roegen N, 1971, *The Entropy Law and the Economic Process*, Harvard University Press, Cambridge, MA.

German Advisory Council on Global Change, 2002, *Charging for the Use of Global Commons*, WBGU, Berlin.

Gertler M S, 1997, 'The invention of regional culture', in Lee R and Wills J (eds), *Geographies of Economies*, Arnold, London, 47–58.

Gertler M S, 2001, 'Best practice? Geography, learning and the institutional limits to strong convergence', *Journal of Economic Geography*, 1, 5–26.

Gertler M S and Di Giovanna S, 1997, 'In search of the new social economy: collaborative relations between users and producers of advanced manufacturing technologies', *Environment and Planning A*, 29,1585–602.

Ghosal S and Bartlett C A, 1997, *The Individualized Corporation*, Heinemann, London.

Giddens A, 1991, *Modernity and Self-Identity: Self and Society in the Late Modern Age*, Polity Press, Cambridge.

Gilroy P, 1993, *The Black Atlantic: Modernity and Double Consciousness*, Verso, London.

Glaeser E L, Kolko J and Saiz A, 2001, 'Consumer city', *Journal of Economic Geography*, 1, 27–50.

Glennie P and Thrift N, 1996a, 'Consumers, identities and consumption spaces in early-modern England', *Environment and Planning A*, 28, 25–45.

Glennie P and Thrift N, 1996b, 'Consumption, shopping and gender', in Wrigley N and Lowe M (eds), *Retailing, Consumption and Capital: Towards a New Retail Geography*, Addison Wesley Longman, Harlow, 221–37.

Goldsmith E, 1992, *The Way: An Ecological Worldview*, University of Georgia Press, Athens, GA.

Goodin R, 1992, *Green Political Theory*, Polity Press, Cambridge.

Goodman D, Sarj J and Wilkinson J, 1987, *From Farming to Biotechnology*, Blackwell, Oxford.

Goodman D and Watts M, 1997, *Globalising Food: Agrarian Questions and Global Restructuring*, Routledge, London.

Goss J, 1992, 'Modernity and post-modernity in the retail landscape', in Anderson K and Gale F (eds), *Inventing Places: Studies in Cultural Geography*, Longman Cheshire, Melbourne, 158–77.

Goss J, 1993, 'The "magic of the mall": an analysis of form, function and meaning in the retail built environment', *Annals of the Association of American Geographers*, 83, 18–47.

Goss J, 1999, 'Once upon a time in the commodity world: an unofficial guide to Mall of America', *Annals of the Association of American Geographers*, 89, 45–75.

Gough J, 2003, 'Review of Producing Places', *Economic Geography*, 79, 96–9.

Grabher G, 2001, 'Ecologies of creativity: the village, the group and the heterarchic organization of the British advertising industry', *Environment and Planning A*, 33, 351–74.

Grabher G, 2002, 'The project ecology of advertising: tasks, talents and teams', *Regional Studies*, 36, 245–62.

Gramsci A, 1971, Selections from the Prison Notebooks, Lawrence and Wishart, London.

Gregson N, 1995, 'And now it's all consumption?', *Progress in Human Geography*, 19, 135–41.

Gregson N and Crewe L, 1994, 'Beyond the high street and the mall: car boot fairs and new geographies of consumption in the 1990', *Area*, 26, 261–7.

Gregson N and Crewe L, 1997, 'The bargain, the knowledge and the spectacle: making sense of consumption in the space of the car boot sale', *Society and Space*, 15, 87–112.

Gregson N, Crewe L and Brooks K, 2001, *Second Hand Worlds*, Routledge, London.

Griffiths S and Wallace J, (eds), 1998, *Consuming Passions: Food in the Age of Anxiety*, Mandolin, Manchester.

Habermas J, 1976, *Legitimation Crisis*, Heinemann, London.

Hadjimichalis C and Vaiou D, 1996, 'Informalisation along global commodity chains: some evidence from southern Europe', Paper presented to the Conference on La Economia Sumergita, Alicante, 15–20 August.

Hagedoon J, 1993, 'Understanding the rationale of strategic technology partnering: interorganizational modes of cooperation and sectoral differences', *Strategic Management Journal*, 14, 371–85.

Hagerstrand T, 1975, 'Space, time and human conditions', in Karlqvist A, Lundquist L and Snickars F (eds), *Dynamic Allocation of Urban Space*, Saxon House, Farnborough, 3–14.

Hagstrom P and Hedlund G, 1998, 'A three dimensional model of changing internal structure in the firm', in Chandler A, Hagstrom P and Solvell U (eds), *The Dynamic Firm: The Role of Technology, Strategy, Organisation and Regions*, Oxford University Press, Oxford, 166–91.

Hajer M, 1995, *The Politics of Environmental Discourse: Ecological Modernisation and the Policy Process*, Clarendon Press, Oxford.

Halford S and Savage M, 1997, 'Rethinking restructuring: embodiment, agency and identity in organisational change', in Lee R and Wills J (eds), *Geographies of Economies*, Arnold, London, 108–17.

Hall S, 1991, 'The local and the global: globalization and ethnicity', in King A D (ed.), *Culture, Globalization and the World System*, Macmillan, London, 19–30.

Hall W, 2001, 'ABB entrusts its future to "brain power"', *Financial Times*, 12 January.

Hamer M, 1974, *Wheels Within Wheels*, Friends of the Earth, London.

Hamilton A, 2003, 'Fashion puts the lid on Tupperware parties', *The Times*, 24 January.

Hamilton N, 1993, 'Who owns dinner? Evolving legal mechanisms for ownership of plant genetic resources', *Tulsa Law Journal*, 26, 587–657.

Hampton M, 1998, 'Backpacker tourism and economic development', *Annals of Tourism Research*, 25, 639–60.

Hannigan J, 1998, *Fantasy City: Pleasure and Profit in the Postmodern Metropolis*, Routledge, London.

Hardy J, 2002, 'An institutionalist analysis, of foreign Investment in Poland: Wroclaw's second great transformation', Unpublished PhD thesis, University of Durham.

Harvey D, 1982, *The Limits to Capital*, Blackwell, Oxford.

Harvey D, 1985, 'The geopolitics of capitalism', in Gregory D and Urry J (eds), *Social Relations and Spatial Structure*, Macmillan, Basingstoke, 128–63.

Harvey D, 1989, *The Condition of Postmodernity*, Blackwell, Oxford.

Harvey D, 1996, *Justice, Nature and the Geography of Difference*, Blackwell, Oxford.

Harvey D, 2002, 'Reflecting on "The Limits to Capital"', Paper presented to the Annual Conference of the Association of American Geographers, Los Angeles, 19–23 March.

Harvey F, 2003, 'Branding: a crucial defence in guarding market share', *Financial Times Business of Chemicals Special Report*, 10 September.

Harvey M, Quilley S and Beynon H, 2003, *Exploring the Tomato: Transformations of Nature, Society and Economy*, Edward Elgar, Cheltenham.

Hay C and Jessop B, 1995, 'Introduction: local political economy: governance and regulation', *Economy and Society*, 24, 303–6.

Heelas P, 2002, 'Work ethic, soft capitalism and the "turn to life"', in Gay P and Pryke M (eds), *Cultural Economy*, Sage, London, 78–96.

Held D, 1989, *Political Theory and the Modern State: Essays on State, Power and Democracy*, Stanford University Press, Stanford, CA.

Hirsch J, 1978, 'The state apparatus and social reproduction: elements of a theory of the bourgeois state', in Holloway J and Picciotto S (eds), *State and Capital: A Marxist Debate*, Arnold, London, 57–107.

Hirst P and Thompson G, 1995, *Globalization in Question*, Polity Press, Cambridge.

Hoagland P, 1993, 'Manganese nodule price trends', *Resource Policy*, 19, 287–98.

Hodgson G, 1988, *Economics and Institutions: A Manifesto for Modern Institutional Economics*, Polity Press, Cambridge.

Hodgson G, 1993, E*conomics and Evolution: Bringing Life Back into Economics*, Polity Press, Cambridge.

Hogarth T and Daniel W W, 1989, *Britain's New Industrial Gypsies*, Policy Studies Institute, London.

Hollingsworth J Rogers, 2000, 'Doing institutional analysis', *Review of International Political Economy*, 7, 595–640.

Holloway J and Picciotto S (eds), 1978, *State and Capital: A Marxist Debate*, Arnold, London.

Hope K, 2003, 'Criminals exploit $2bn business', *Financial Times*, 11 November.

Hopkins J, 1990, 'West Edmonton Mall: landscape of myths and elsewhereness', *Canadian Geographer*, 34, 2–17.

Houlder V, 2002, 'Councils clash with electrical goods industry over waste laws', *Financial Times*, 28/29 December.

Howells J R, 1993, 'Emerging global strategies in innovation management', in Humbert M (ed.), *The Impact of Globalisation on Europe's Firms and Industries*, Pinter, London, 219–28.

Hudson M, 2001, 'Re-presenting a region: the social construction of a regional history', Unpublished MSc thesis, Department of Geography, University of Bristol.

Hudson R, 1983, 'Capital accumulation and chemicals production in Western Europe in the postwar period', *Environment and Planning A*, 15, 105–22.

Hudson R, 1986, 'Nationalised industry policies and regional policies: the role of the state in the deindustrialisation and reindustrialisation of regions', *Society and Space,* 4, 7–28.

Hudson R, 1989, 'Labour market changes and new forms of work in old industrial regions: maybe flexibility for some but not flexible accumulation', *Society and Space*, 7, 5–30.

Hudson R, 1994a, 'New production concepts, new production geographies? Reflections on changes in the automobile industry', *Transactions of the Institute of British Geographers*, 19, 331–45.

Hudson R, 1994b, 'Institutional change, cultural transformation and economic regeneration: myths and realities from Europe's old industrial regions', in Amin A and Thrift N (eds), *Globalization, Institutions and Regional Development in Europe*, Oxford University Press, Oxford, 331–45.

Hudson R, 1994c, 'Restructuring production in the West European steel industry', *Tijdschrift voor Economische en Sociale Geografie*, 85, 99–113.

Hudson R, 1995, 'The Japanese, the European market and the automobile industry in the United Kingdom', in Hudson R and Schamp E W (eds), *Towards a New Map of Automobile Manufacturing in Europe? New Production Concepts and Spatial Restructuring*, Springer, Berlin, 63–92.

Hudson R, 1997, 'The end of mass production and of the mass collective worker? Experimenting with production, employment and their geographies', in Lee R and Wills J (eds), *Geographies of Economies*, Arnold, London, 302–10.

Hudson R, 1999, 'The learning economy, the learning firm and the learning region: a sympathetic critique of the limits to learning', *European Urban and Regional Studies*, 6, 59–72.

Hudson R, 2001, *Producing Places*, Guilford, New York.

Hudson R, 2002, 'Changing industrial production systems and regional development in the New Europe', *Transactions, Institute of British Geographers*, 27, 262–81.

Hudson R, 2003, 'Global production systems and European integration', in Peck J and Yeung H W-C (eds), *Global Connections*, Sage, London, 216–30.

Hudson R and Lewis J, 1985, 'Introduction: recent economic, social and political changes in southern Europe', in Hudson R and Lewis J (eds), *Uneven Development in Southern Europe*, Methuen, London, 1–53.

Hudson R and Schamp E W, 1995, 'Interdependent and uneven development in the spatial reorganisation of the automobile production systems in Europe', in Hudson R and Schamp E W (eds), *Towards a New Map of Automobile Manufacturing in Europe? New Production Concepts and Spatial Restructuring*, Springer, Berlin, 219–44.

Hudson R and Weaver P, 1997, 'In search of employment creation via environmental valorisation: exploring a possible Eco-Keynesian future for Europe', *Environment and Planning*, 29, 1647–61.

Hughes A, 2000, 'Retailers, knowledges and changing commodity networks: the case of the cut flower trade', *Geoforum*, 31, 175–90.

International Organisation for Migration, 2003, *Annual World Migration Report*, available at http://www.iom.int [accessed 11 June 2003].

Jackson P, 1993, 'Towards a cultural politics of consumption', in Bird J, Curtis B, Putnam T, Robinson G and Tickner L (eds), *Mapping the Futures: Local Cultures, Global Changes*, Routledge, London, 207–28.

Jackson P, 2002, 'Commercial cultures: transcending the cultural and the economic', *Progress in Human Geography*, 26, 3–18.

Jackson P and Taylor J, 1996, 'Geography and the cultural politics of advertising', *Progress in Human Geography*, 20, 356–71.

Jackson P and Thrift N, 1995, 'Geographies of consumption', in Miller D (ed.), *Acknowledging Consumption*, Routledge, London, 204–37.

Jackson T, 1995, *Material Concerns*, Routledge, London.

Jameson F, 1988, *The Ideologies of Theory, (vol. 2)*, Routledge, London.

Jenson J, 1990, 'Representations in crisis: the root of Canada's permeable Fordism', *Canadian Journal of Political Science*, 23, 653–83.

Jessop B, 1982, *The Capitalist State*, Martin Robertson, Oxford.

Jessop B, 1990, *State Theory: Putting Capitalist States in Their Place*, Cambridge University Press, Cambridge.

Jessop B, 1997, 'Capitalism and its future: remarks on regulation, government and governance', *Review of International Political Economy*, 4, 561–81.

Jessop B, 2000, 'The state and the contradictions of the knowledge-driven economy', available at http://www.comp.lancs.ac.uk/sociology/soc044rj.html [accessed 23 October 2001].

Jessop B, 2001, 'Institutional (re)turns and the strategic-relational approach', *Environment and Planning A*, 33, 1213–35.

Jessop B, Bonnet K, Bromley S and Ling T, 1988, *Thatcherism: A Tale of Two Nations*, Polity Press, Cambridge.

Johnson R, 1986, 'The story so far: and other transformations?', in Punter D (ed.), *Introduction to Contemporary Cultural Studies*, Longman, London, 277–313.

Jones M, 1997, 'Spatial selectivity of the state: the regulationist enigma and local struggles over economic governance', *Environment and Planning A*, 29, 831–64.

Kenney M and Curry J, 2001, 'Beyond transaction costs: e-commerce and the power of the Internet dataspace', in Leinbach T R and Brunn S D (eds), *Worlds of E-Commerce*, Wiley, Chichester, 45–66.

King R, 1995, 'Migrations, globalisation and place', in Massey D and Jess P (eds), *A Place in the World? Places, Cultures and Globalization*, Oxford University Press, Oxford, 5–44.

King R, Warnes A W and Williams A M, 2000, *Sunset Lives: British Retirement Migration to the Mediterranean*, Berg, London.

Kitchen R, 1998, *Cyberspace*, Wiley, Chichester.

Klein N, 2000, *No Logo*, Harper Collins, London.

Kloppenburg J, 1988, *First the Seed. The Political Economy of Plant Biotechnology, 1492–2000*, Cambridge University Press, Cambridge.

Kowinski W S, 1985, *The Malling of America: An Inside Look at the Great Consumer Paradise*, Morrow, New York.

Krugman P, 2000, 'Where in the world is the "New Economic Geography"', in Clark G L, Feldman M P and Gertler M (eds), *The Oxford Handbook of Economic Geography*, Oxford University Press, Oxford, 49–60.

Lakka S, 1994, 'The new international division of labour and the Indian software industry', *Modern Asian Studies*, 28, 381–408.

Lane C and Reinhard B (eds), 1998, *Trust Within and Between Organisations*, Oxford University Press, Oxford.

Lappé F M, 1991, *Diet for a Small Planet*, Ballantine, New York.

Lash S and Urry J, 1987, *The End of Organised Capitalism*, Polity Press, Cambridge.

Lash S and Urry J, 1994, *Economies of Signs and Space*, Sage, London.

Lash S and Friedman J (eds), 1992, *Modernity and Identity*, Blackwell, Oxford.

Latour B, 1987, *Science in Action: How to Follow Scientists and Engineers through Society*, Open University Press, Milton Keynes.

Law J, 2002, 'Economics as interference', in Du Gay P and Pryke M (eds), *Cultural Economy*, Sage, London, 21–38.

Leborgne D and Lipietz A, 1991, 'Two social strategies in the production of new industrial spaces', in Benko G and Dunford M (eds), *Industrial Change and Regional Development: The Transformation of New Industrial Spaces*, Belhaven, London, 27–50.

Lee R, 2002, '"Nice maps, shame about the theory"? Thinking geographically about the economic', *Progress in Human Geography*, 26, 3, 333–54.

Leidner R, 1993, *Fast Food and Fast Talk: Service Work and the Routinization of Everyday Life*, University of California Press, Berkeley, CA.

Leinbach T R, 2001, 'Emergence of the digital economy and e-commerce', in Leinbach T R and Brunn S D (eds), *Worlds of E-Commerce*, Wiley, Chichester, 3–26.

Leinbach T R and Brunn S D (eds), 2001, *Worlds of E-Commerce*, Wiley, Chichester.

Leonard D and Swap W, 1999, *When Sparks Fly: Igniting Creativity in Groups*, Harvard Business School Press, Boston, MA.

Leonard H J, 1988, *Pollution and the Struggle for the World Product: Multinational Corporations, Environment and International Competitive Advantage*, Cambridge University Press, Cambridge.

Leontidou L, 1993, 'Informal strategies of unemployment relief in Greek cities: the relevance of family, locality and housing', *European Planning Studies* 1, 43–68.

Leslie D, 1994, 'Global scan: the globalisation of advertising agencies, concepts and campaigns', *Economic Geography*, 71, 402–26.

Leslie, D. and Butz D, 1998, '"GM suicide": flexibility, space and the injured body', *Economic Geography*, 74, 360–78.

Leslie D and Reimer S, 1999, 'Spatializing commodity chains', *Progress in Human Geography*, 23, 401–20.

Lester T, 1998, 'Electric effect of alliances', *Financial Times*, 15 January.

Levinthal D, 1996, 'Learning and Schumpeterian dynamics', in Dosi G and Malerba F (eds), *Organisation and Strategy in the Evolution of the Enterprise*, Macmillan, London, 27–41.

Leyshon A, 2000, 'Money and finance', in Sheppard E and Barnes T (eds), *A Companion to Economic Geography*, Blackwell, Oxford, 432–49.

Leyshon A, Lee R and Williams C C (eds), 2003, *Alternative Economic Spaces*, Sage, London.

Leyshon A and Thrift N, 1992, 'Liberalization and consolidation: the Single European Market and the remaking of European financial capita', *Environment and Planning A*, 24, 49–81.

Lie T C and Santucci G, 1993, 'Seeking balanced trade and competition in the context of globalisation: the case of electronics', in Humbert M (ed.), *The Impact of Globalisation on Europe's Firms and Regions*, Pinter, London, 114–24.

Lipietz A, 1986, 'New tendencies in the international division of labour: regimes of accumulation and modes of regulation', in Scott A J and Storper M (eds), *Production, Territory, Work*, Unwin Hyman, London, 16–40.

Lipietz A, 1987, *Mirages and Miracles*, Verso, London.

Lofgren O, 1990, 'Consuming interests', *Culture and History*, 7, 7–36.

Lorange P and Roos J, 1993, *Strategic Alliances*, Blackwell, Oxford.

Lotz J, 1998, *The Lichen Factor: The Search for Community Development in Canada*, UCCB Press, Sydney.

Lovelock J, 1988, *The Ages of Gaia: A Biography of Our Living Earth*, Oxford University Press, Oxford.

Lowe M and Wrigley N, 1996, 'Towards a new retail geography', in Wrigley N and Lowe M (eds), *Retailing, Consumption and Capital: Towards a New Retail Geography*, Addison Wesley Longman, Harlow, 3–30.

Lucas R E B, 2001, 'The effects of proximity and transportation on developing country population migrations', *Journal of Economic Geography*, 1, 323–40.

Lundvall B A, 1992, *National Systems of Innovation: Towards a Theory of Innovation and Interactive Learning*, Pinter, London.

Lundvall B A, 1995, 'The learning economy – challenges to economic theory and policy', Revised paper originally presented to the European Association of Evolutionary Political Economists, Copenhagen, 27–29 October 1994.

Lundvall B A and Johnson B, 1994, 'The learning economy', *Journal of Industry Studies*, 2, 23–42.

Lury C, 2000, 'The united colors of diversity', in Franklin S, Lury C and Stacey J (eds), *Global Nature, Global Culture*, Sage, London, 146–87.

Luseby J, 1998, 'Smaller, cheaper and safer', *Financial Times*, 8 September.

Lutz H, 2002, 'At your service madam! The globalisation of domestic service', *Feminist Review*, 70, 89–104.

MacKinnon D, 2000, 'Managerialism, governmentality and the state: a neo-Foucauldian approach to local economic governance', *Political Geography*, 19, 293–314.

Maclaughlan J and Salt J 2002, *Migration Policies Towards Highly Skilled Foreign Workers*, Home Office, London.

MacLeod G, 2003, *Privatising the City? Edge Cities and Gated Communities in the United Kingdom: A Report to the Office of the Deputy Prime Minister*, International Centre for Regional Regeneration and Development Studies, Wolfson Research Institute, University of Durham, 24 pp.

MacLeod G and Jones M 1998, 'Re-regulating a region? Institutional fixes, entre-preneurial, discourse and the politics of representation', University of Wales, Aberystwyth and University of Manchester, mimeo.

Maffesoli M, 1991, 'The ethics of aesthetics', *Theory, Culture and Society*, 8, 7–20.

Maffesoli M, 1992, *The Time of the Tribes*, Sage, London.

Mahroum S, 1999, 'Highly skilled globetrotters', in OECD (ed.), *Mobilising Human Resources for Innovation. Proceedings of the OECD Workshop on Science and Technology*, OECD, Paris.

Manzagol C, 1991, 'The rise of a technological complex: some comments on the Phoenix case', in Benko G and Dunford M (eds), *Industrial Change and Regional Development: The Transformation of New Industrial Spaces*, Belhaven, London, 237–49.

Marples D R, 1987, *Chernobyl and Nuclear Power in the USSR*, Macmillan, London.

Marsden T, 1998, 'Creating competitive space: exploring the social and political maintenance of natural power', *Environment and Planning A*, 30, 481–98.

Marsden T, Flynn A and Harrison M, 2000, *Consuming Interests: The Social Provision of Foods*, UCL Press, London.

Marshall G, 1986, 'The workplace culture of a licensed restaurant', *Theory, Culture and Society*, 3, 33–48.

Martin R, 1994, 'Stateless monies, global financial integration and national economic autonomy: the end of geography?', in Corbridge S, Martin R and Thrift N (eds), *Money, Power, Space*, Blackwell, Oxford, 253–78.

Maskell P, Eskelinen H, Hannibalsson I, Malmberg A and Vatne E, 1998, *Competitiveness, Localised Learning and Regional Development*, Routledge, London.

Massey D, 1995, *Spatial Divisions of Labour: Social Structures and the Geography of Production*, Macmillan, London.

Massey D and Catalano A, 1978, *Capital and Land: Landownership by Capital in Britain*, Arnold, London.

Massey D, Quintas P and Wield D, 1992, *High Tech Fantasies: Science Parks in Society, Science and Space*, Routledge, London (2nd edition).

Mattick P, 1971, *Marx and Keynes: The Limits of the Mixed Economy*, Merlin, London.

McCracken G, 1988, *Culture and Consumption*, Indiana University Press, Bloomington, IN.

McDowell L, 1994, 'The transformation of cultural geography', in Gregory D, Martin R and Smith D (eds), *Human Geography: Society, Space and Social Science*, Macmillan, Basingstoke, 146–73.

McDowell L, 1997, 'A tale of two cities? Embedded organisations and embodied workers in the City of London', in Lee R and Wills J (eds), *Geographies of Economies*, Arnold, London, 118–29.

McDowell L, 2001, 'Linking scales: or how research about gender and organization raises new issues for economic geography', *Journal of Economic Geography*, 1, 227–50.

McFall L, 2002, 'Advertising, persuasion and the culture/economy dualism', in Du Gay P and Pryke M (eds), *Cultural Economy*, Sage, London, 148–65.

McGoldrick P, 1991, *Retail Marketing*, McGraw-Hill, Maidenhead.

McGrew A, 1995, 'World order and political space', in Anderson J, Brook C and Cochrane A (eds), *A Global World?* Oxford University Press, Oxford, 11–64.

McManus P, 1996, 'Contested terrains: politics, stories and discourses of sustainability', *Environmental Politics*, 5, 48–73.

McRae H, 1997, 'My Office? No, its a white collar factory', *Independent on Sunday*, 7 December.

McRobbie A, 1997, 'Bridging the gap: feminism, fashion and consumption', *Feminist Review*, 55, 73–89.

McRobbie A, 2002, 'From Holloway to Hollywood: happiness at work in the new cultural economy', in Du Gay P and Pryke M (eds), *Cultural Economy*, Sage, London, 97–114.

Meadows D H, Meadows D L and Randers J, 1972, *The Limits to Growth: A Report for the Club of Rome's Project on the Predicament of Mankind*, Earth Island, London.

Metcalfe J S, 1995, 'Technology system and technology policy in an evolutionary framework', *Cambridge Journal of Economics* 19, 25–46.

Metcalfe J S and James A, 2000, 'Emergent innovation systems and the delivery of clinical services: the case of intracocular lenses', *Working Paper Number 9*, Centre for Research on Innovation and Competition, University of Manchester.

Michalowski R J and Kramer R C, 1987, 'The space between laws: the problem of corporate crime in a transnational context', *Social Problems*, 34, 34–53.

Miller D, 1987, *Material Culture and Mass Consumption*, Blackwell, Oxford.

Miller D, 2002, 'The unintended political economy', in Du Gay P and Pryke M (eds), *Cultural Economy*, Sage, London, 166–84.

Miller D, Jackson P, Thrift N, Holbrook B and Rowlands M, 1998, *Shopping, Place and Identity*, Routledge, London.

Miller D and Rose N, 1997, 'Mobilising the consumer: assembling the subject of consumption', *Theory, Culture and Society*, 14, 1–36.

Miller P, Pons J N and Naude P, 1996, 'Global teams', *Financial Times*, 14 June.

Miller R, 1991, '"Selling Mrs Consumer": advertising and the creation of suburban socio-spatial relations', *Antipode*, 23, 263–301.

Minton A, 2000, 'Suppliers sue M&S for £53.6m', *Financial Times*, 11 January.

Minton A, 2002, *Building Balanced Communities: The US and UK Compared*, Royal Institute of Chartered Surveyors, London.

Mitchell A, 1995, 'Un-American activities', *London Evening Standard*, 13 September.

Mitchell P, 1998, 'Innovation in the right place', *Financial Times*, 17 April.

Morgan K, 1995, 'The learning region: institutions, innovation and regional renewal', *Papers in Planning Research No. 15*, Department of City and Regional Planning, Cardiff University.

Morgan K and Murdoch J, 2000, 'Organic versus conventional agriculture: know-ledge, power and innovation in the food chain', *Geoforum*, 31, 159–73.

Morris D and Hergert M, 1987, 'Trends in international collaborative agreements', *Columbia Journal of World Business*, XXII, 15–21.

Morrison S, 2003, 'Silicon Valley fumes as jobs migrate to foreign fields', *Financial Times*, 24 September.

Mort F, 1996, *Cultures of Consumption: Masculinities and Social Space in Late Twentieth-century Britain*, Routledge, London.

Mort F, 1997, 'The coming of the mass market', in Nava M (ed.), *Buy This Book: Studies in Advertising and Consumption*, Routledge, London.

Mulgan G, 1991, *Communication and Control: Networks and New Economics of Communication*, Polity Press, Cambridge.

Murdoch J, Marsden T and Banks J, 2000, 'Quality, nature and embeddedness: some theoretical considerations in the context of the food sector', *Economic Geography*, 76, 107–25.

Nava M, 1997, 'Modernity's disavowal: women, the city and the department store', in Falk P and Campbell C (eds), *The Shopping Experience*, Sage, London, 56–92.

Negrin L, 1999, 'The self as image: a critical appraisal of postmodern theories of fashion', *Theory, Culture and Society*, 16, 99–120.

Nitzan J, 2001, 'Regimes of differential accumulation: mergers, stagflation, and the logic of globalisation', *Review of International Political Economy*, 8, 226–74.

Nixon S, 1996, *Hard Looks: Masculinities, Spectatorship and Contemporary Consumption*, UCL Press, London.

Nixon S, 2002, 'Re-imagining the ad agency: the cultural connotations of economic forms', in Du Gay P and Pryke M (eds), *Cultural Economy*, Sage, London, 132–47.

Nonaka I and Takeuchi H, 1995, *The Knowledge-Creating Company*, Oxford University Press, New York.

Nonaka I, Toyama R and Konno U, 2001, 'SECI, *Ba* and leadership: a unified model of dynamic knowledge creation', in Nonaka I and Teece D (eds), *Managing Industrial Knowledge: Creation, Transfer, Utilization*, Sage, London, 68–90.

Noteboom B, 1999, 'Innovation, learning and industrial organisations', *Cambridge Journal of Economics*, 23, 127–50.

O'Connor G, 2000, 'Mining giants take control of merger wave of $1bn deals', *Financial Times*, 6 September.

O'Connor J, 1994, 'Is sustainable capitalism possible?', in O'Connor M (ed.), *Is Capitalism Sustainable?*, Guilford, New York, 152–75.

O'Connor M, 1994a, 'Co-dependency and indeterminacy: a critique of the theory of production', in O'Connor M (ed.), *Is Capitalism Sustainable?* Guilford, New York, 51–75.

O'Connor M, 1994b, 'On the misadventures of capitalist nature', in O'Connor M (ed.), *Is Capitalism Sustainable?* Guilford, New York, 126–44.

Offe C, 1975, 'The theory of the capitalist state and the problem of policy formation', in Lindberg L N, Alford R, Crouch C and Offe C (eds), *Stress and Contradiction in Modern Capitalism*, DC Heath, Lexington, MA, 125–44.

Offe C, 1976, 'Political authority and class structures', in Connerton P (ed.), *Critical Sociology*, Penguin, Harmondsworth, 388–421.

Ohmae K, 1995, *The End of the Nation State*, Free Press, Glencoe, IL.

Okamura C and Kawahito H, 1990, *Karoshi*, Mado Sha, Tokyo.

O'Neill P, 1997, 'Bringing the qualitative state into economic geography', in Lee R and Wills J (eds), *Geographies of Economies*, Arnold, London, 290–301.

O'Riordan T, 1981, *Environmentalism*, Pion, London.

Painter J and Goodwin M, 1995, 'Local governance and concrete research: investigating the uneven development of regulation', *Economy and Society*, 24, 334–56.

Paley W S, 1952, *Resources for Freedom*, A Report to the President by the Provident's Materials Policy Commission, US Government Printing Office, Washington, DC.

Palloix C, 1976, 'The labour process from Fordism to Neo-Fordism', in *Conference of Socialist Economists: The Labour Process and Class Strategies*, London, 46–67.

Palloix C, 1977, 'The self-expansion of capital on a world scale', *Review of Radical Political Economics*, 9,1–28.

Panitch L, 1996, 'Re-thinking the role of the state', in Mittleman J H (ed.), *Globalization: Critical Reflections*, Lynne Rienner, Boulder, CO, 83–113.

Pasternack B A and Viscio A J, 1998, *The Centerless Corporation: A New Model for Transforming your Organisation for Growth and Prosperity*, Simon & Schuster, New York.

Pearce D, Markandya A and Barbier E, 1989, *Blueprint for a Green Economy*, Earthscan, London.

Pearson R and Mitter S, 1994, 'Employment and working conditions of low-skilled information-processing workers in less developed countries', *International Labour Review*, 132, 49–64.

Peck J, 1994, 'Regulating labour: the social regulation and reproduction of local labour markets', in Amin A and Thrift N J (eds), *Globalization, Institutions and Regional Development in Europe*, Oxford University Press, Oxford, 147–76.

Peck J, 1995, *Workplace: The Social Regulation of Labour Markets*, Guilford, New York.

Peck J, 2004, 'Political economies of scale: fast policy, interscalar relations and neoliberal workfare', *Economic Geography* (forthcoming).

Peck J and Theodore N C, 1997, 'Trading warm bodies: processing contingent labour in Chicago's temp industry', International Centre for Labour Studies, University of Manchester, mimeo, 24 pp.

Penrose E, 1957, *The Theory of the Growth of the Firm*, Blackwell, Oxford.

Pepper D, 1984, *Roots of Modern Environmentalism*, Routledge, London.

Pickard J, 2002, 'Records broken as commuters take to travelling new lengths', *Financial Times*, 3 September.

Pilling D, 2000, 'Glaxo, SB to announce deal today', *Financial Times*, 17 January.

Pine B J, 1993, *Mass Customization: The New Frontier in Business Competition*, Harvard University Press, Cambridge, MA.

Pollan M, 1999, 'In my own backyard', *Independent on Sunday: Sunday Review*, 4 April.

Pollert A, 1988, 'Dismantling flexibility', *Capital and Class*, 34, 42–75.

Poon A, 1989, 'Competitive strategies for a "new tourism"', in Cooper C (ed.), *Progress in Tourism, Recreation and Hospitality Management*, Belhaven, London, 91–102.

Portes A, Castells M and Benton L A, 1988, *The Informal Economy: Studies in Advanced and Less Developed Countries*, Johns Hopkins University Press, Baltimore, MD.

Postone M, 1996, *Time, Labour and Social Domination*, Cambridge University Press, Cambridge.

Pred A, 1996, 'Interfusions: consumption, identity and the practices and power relations of everyday life', *Environment and Planning A*, 28, 11–24.

Pretty J , 2001, 'Against the grain', *The Guardian*, 17 January.

Probyn E, 1998, 'Mc-identities: food and the familial citizen', *Theory, Culture and Society*, 15, 155–73.

Radice H, 2000, 'Globalization and national capitalism: theorizing convergence and differentiation', *Review of International Political Economy*, 7, 19–42.

RAFI (Rural Advancement Foundation International), 1999, 'The gene giants: up-date on consolidation in the life industry', *Rafi communique*, 30 March.

Ramsay H, 1992, 'Whose champions? Multinationals, labour and industry policy in the European Community after 1992', *Capital and Class*, 48, 17–40.

Rankin S, 1987, 'Exploitation and the labour theory of value: a neo-Marxian reply', *Capital and Class*, 32, 104–16.

Ray L and Sayer A, 1999, 'Introduction', in Ray L and Sayer A (eds), *Culture and Economy after the Cultural Turn*, Sage, London, 1–16.

Redclift N and Mingione E (eds), 1986, *Beyond Employment: Household, Gender and Subsistence*, Blackwell, Oxford.

Roberts B, 1987, 'Marx after Steedman – separating Marxism and "surplus theory"', *Capital and Class*, 32, 84–103.

Robins K, 1989, 'Global times', *Marxism Today*, 20–7 December,.

Rose N, 1996, 'The death of the social? Re-figuring the territory of government', *Economy and Society*, 25, 327–56.

Rose N, 1999, *Powers of Freedom: Reframing Political Thought*, Cambridge University Press, Cambridge.

Rose N and Miller P, 1992, 'Political power beyond the state: problematics of government', *British Journal of Sociology*, 42, 202–23.

Rosset P and Benjamin M (eds), 1994, *The Greening of the Revolution: Cuba's Experiment with Organic Agriculture*, Ocean Press, Melbourne.

Ruggie J G, 1993, 'Territoriality and beyond: problematizing modernity in international relations', *International Organization*, 27, 139–74.

Sack R, 1992, *Place, Modernity and the Consumer's World*, Johns Hopkins University Press, Baltimore, MD.

Sampson H, 2003, 'Transformational drifters or hyperspace dwellers? An exploration of the lives of Filipino seafarers aboard and ashore', *Ethnic and Racial Studies*, 26, 253–77.

Sassen S, 1991, *The Global City: New York, London, Tokyo*, University of California Press, Los Angeles, CA.

Sassen S, 2003, 'Globalization or denationalization?', *Review of International Political Economy*, 10, 1–22.

Saxenian A, 2000, 'Silicon Valley's new immigrant entrepreneurs', *Working Paper No. 15*, Centre for Comparative Immigration Studies, University of California, San Diego, CA.

Sayer A, 1984, *Method in Social Science*, Hutchinson, London.

Sayer A, 1989, 'Post-Fordism in question', *International Journal of Urban and Regional Research*, 13, 666–95.

Sayer A and Walker R, 1992, *The New Social Economy: Reworking the Division of Labour*, Blackwell, Oxford.

Schamp E W, 1991, 'Towards a spatial reorganisation of the German car industry? The implications of new production concepts', in Benko G and Dunford M (eds), *Industrial Change and Regional Development: The Transformation of New Industrial Spaces*, Belhaven, London, 159–70.

Scharb M, 2001, 'Eco-industrial development: a strategy for building sustainable communities', *Review of Economic Development Interaction and Practice 8*, Cornell University and US Economic Development Administration, 43 pp.

Scharmer C O, 2001, 'Self-transcending knowledge: organizing around emerging realities', in Nonaka I and Teece D (eds), *Managing Industrial Knowledge: Creation, Transfer, Utilization*, Sage, London, 68–90.

Schlereth T J, 1989, 'Country stores, country fairs and mail order catalogues: consumption in rural America', in Bronner S J (ed.), *Consuming Visions: Accumulation and Display of Goods in America 1880–1920*, Norton, New York, 339–75.

Schmidt V A, 2002, *The Futures of European Capitalism*, Oxford University Press, Oxford.

Seager J, 1995, *The State of the Environment Atlas*, Penguin, Harmondsworth.

Shaw G and Williams A, 1994, *Critical issues in Tourism*, Blackwell, Oxford.

Sheller M and Urry J, 2000, 'The city and the car', *International Journal of Urban and Regional Research*, 24, 737–57.

Sherman B, 1985, *The New Revolution*, Wiley, Chichester.

Shields R, 1989, 'Social spatialization and the built environment: the West Edmonton Mall', *Society and Space*, 7, 147–64.

Shiva V, 1997, *Biopiracy: The Plunder of Nature and Knowledge*, Southend Press, Boston, MA.

Showalter E, 2001, 'Fade to greige', *London Review of Books*, 23, 14 January, 1–9.

Sirpa T, 2002, 'Whose place is this space? Life in the street prostitution area of Helsinki, Finland', *International Journal of Urban and Regional Research*, 26, 210–28.

Slater D, 2002, 'Capturing markets from the economists', in Du Gay P and Pryke M (eds), *Cultural Economy,* Sage, London, 59–77.

Smith A, 1998, *Reconstructing the Regional Economy: Industrial Transformation and Regional Development in Slovakia*, Edward Elgar, Cheltenham.

Smith A, Rainnie A, Dunford M, Hardy J, Hudson R and Sadler D, 2002, 'Networks of value, commodities and regions: reworking divisions of labour in macro-regional economies', *Progress in Human Geography*, 26, 41–64.

Smith N, 1984, *Uneven Development: Nature, Capital and the Production of Space*, Blackwell, Oxford.

Smith S, 2003, 'Why (some) housing markets work', Seminar presented to the Department of Geography, University of Durham, 18 May.

Soja E W, 2003, 'Seeking spatial justice and global democracy', Paper presented to the International Conference, Rethinking Radical Spatial Approaches, Seminars of the Aegean, Naxos, 8–14 September.

Spohn S G, 2002, 'Eco-industrial parks offer sustainable base redevelopment', available at Http://www.smartgrwoth.org/casestudies/spohn_icma.html [accessed 3 December 2002].

Springett D, 2003, 'Business Conceptions of Sustainable Development', Unpublished PhD thesis, University of Durham.

Stalk G, 1988, 'Time: the next resource of competitive advantage', *Harvard Business Review*, 66, 45–51.

Stockdale M, 2001, 'Chasing those brands on the run', *Financial Times*, 4 December.

Stone C, 2002, 'Environmental consequences of heavy-industry restructuring and economic regeneration through industrial ecology', *Transactions of the Institute of Mining and Metallurgy*, 111, A187–91.

Storper M, 1993, 'Regional "worlds" of production: learning and innovation in the technology districts of France, Italy and the USA', *Regional Studies*, 27, 433–56.

Storper M, 1995, 'The resurgence of regional economies, ten years later: the region as a nexus of untraded interdependencies', *European Urban and Regional Studies*, 2, 191–222.

Strathern M, 1988, *The Gender of the Gift*, University of California Press, Berkeley, CA.

Summers D, 1998, 'Ways to a working woman's heart', *Financial Times*, 20 March.

Tatko J and Robinson T, 2002, 'The Northern Fleet Radiocarbon Waste Development: overview', available at www.nit.org [accessed 18 June 2002].

Taylor M J, 1995, 'Linking economy, environment and policy', in Taylor M J (ed.), *Environmental Change: Industry, Power and Policy*, Avebury, Aldershot, 1–14.

Taylor W, 1991, 'The logic of global business: an interview with ABB's Percy Barnevik', *Harvard Business Review*, March/April.

Thompson E P, 1969, *The Making of the English Working Class*, Penguin, Harmondsworth.

Thompson S, 1999, 'Takeovers, joint ventures and the acquisition of resources for diversification', *Scottish Journal of Political Economy*, 46, 303–18.

Thrift N, 1994, 'On the social and cultural determinants of international financial centres: the case of the City of London', in Corbridge S, Martin R and Thrift N (eds), *Money, Space and Power*, Blackwell, Oxford, 327–55.

Thrift N, 1998, 'The rise of soft capitalism', in Herod A, O'Tuathail G and Roberts S M (eds), *Unruly World: Globalisation, Governance and Geography*, Routledge, London, 25–71.

Thrift N, 1999, 'The globalisation of the system of business knowledge', in Olds K, Dicken P, Kelly P, Kong L and Yeung H W-C (eds), *Globalization and the Asia Pacific*, Routledge, London, 57–71.

Thrift N, 2000, 'Performing cultures in the new economy', *Annals of the Association of American Geographers*, 90, 674–92.

Thrift N, 2001, '"It's the romance, not the finance, that makes the business worth pursuing": disclosing a new market culture', *Economy and Society*, 40, 412–32.

Thrift N, 2002, 'Performing culture in the new economy', in Du Gay P and Pryke M (eds), *Cultural Economy*, Sage, London, 201–34.

Tokar B, 1998, 'Monsanto: a chequered history', *The Ecologist*, 28, 254–61.

Tseng Y-F, 2000, 'The mobility of entrepreneurs and of capital: Taiwanese capital-linked migration', *International Migration*, 38, 143–67.

Turner D, 2003, 'Millions of British posts seem ripe for export as companies look to cut costs', *Financial Times*, 18/19 October.

United Nations – Habitat, 2003, *The Challenge of Slums: Global Report on Human Settlement*, Earthscan, London.

United Nations World Commission on the Environment and Development, 1987, *Our Common Future*, Oxford University Press, Oxford.

Urry J, 1990, T*he Tourist Gaze*, Sage, London.

Urry J, 1999, 'Automobility: car culture and weightless travel – a discussion paper', available at http://www.comp.lancs.ac.uk/sociology/soc008ju.html [accessed 14 August 2002].

Urry J, 2000a, *Sociology Beyond Societies: Mobilities for the Twenty-first Century*, Routledge, London.

Urry J, 2000b, 'Time, complexity and the global', available at http://www.comp.lancs.ac.uk/sociology/soc030ju.html [accessed 14 August 2002]

Urry J, 2001, 'Globalising the tourist gaze', available at http://www.comp.lancs.ac.uk/sociology/soc079ju.html [accessed 14 August 2002].

Vaughan P, 1996, 'Procurement and capital projects', Paper presented to the Conference on Supply Chain Management – Challenges for the 21st century, University of Durham Business School, 9–10 May.

Veltz P, 1991, 'New models of production organisation and trends in spatial development', in Benko G and Dunford M (eds), *Industrial Change and Regional Development: The Transformation of New Industrial Spaces*, Belhaven, London, 193–204.

Virilio P, 1991, *The Lost Dimension*, Semiotext(e), New York.

Vogler J, 1995, *The Global Commons: A Regime Analysis*, Wiley, Chichester.

Von Hippel E, 1998, *The Sources of Innovation*, Oxford University Press, Oxford.

Wagstyl S, 2004 'Borders come down, but what barriers still remain in Europe', *Financial Times*, 14 April.

Walker R, 1985, 'Is there a service economy? The changing capitalist division of labour', *Science and Society*, 49, 42–83.

Walsh V and Galimberti I, 1993, 'Firm strategies, globalisation and new technological paradigms: the case of biotechnology', in Humbert M (ed.), *Impact of Globalisation on Europe's Firms and Regions*, Pinter, London, 175–90.

Warde A, 2002, 'Production, consumption and "cultural economy"', in Du Gay P and Pryke M (eds), *Cultural Economy*, Sage, London, 185–200.

Warf B and Purcell D, 2001, 'The currency of currency: speed, sovereignty and electronic finance', in Leinbach T R and Brunn S D (eds), *Worlds of E-Commerce*, Wiley, Chichester, 223–40.

Waters R and Corrigan C, 1998, 'The cult of gigantism', *Financial Times*, 11/12 April.

Weaver P M, Jansen L, van Grootveld G, van Spiegel E and Vergragt P, 2000, *Sustainable Technology Development*, Greenleaf, Sheffield.

Webster B, 2003, 'Commuters face rail fare free-for-all', *The Times*, 9 May.

Weiss L, 1997, 'Globalisation and the myth of the powerless state', *New Left Review*, 225, 3–27.

Wenger E, 1998, *Communities of Practice: Learning, Meaning and Identity*, Cambridge University Press, Cambridge.

Wernick I D, Herman R, Govinch B and Ausubel J H, 1997, 'Materialization and dematerialization: measures and trends', in Ausubel J H and Langford H (eds), *Technological Trajectories and the Human Environment*, National Academy Press, Washington, DC, 135–56.

Westwood A, 2001, *Not Very Qualified: Raising Skill Levels in the UK Workforce*, The Industrial Society, London.

Whatmore S, 1995, 'From farming to agribusinesses: the global agro-food system', in Johnston R, Taylor P and Watts M (eds), *Geographies of Global Change*, Blackwell, Oxford, 36–49.

Whatmore S, 2002, *Hybrid Geographies*, Sage, London.

Williams A M, Balaz V and Wallace C, 2004, 'International labour mobility and uneven development in Europe', *European Urban and Regional Studies* (forthcoming).

Williams A M and Hall M, 2003, 'Tourism, migration, circulation and mobility', University of Exeter, *mimeo*, 52 pp.

Williams C C and Windebank J, 1998, *Informal Employment in the Advanced Economies*, Routledge, London.

Williams R, 1980, *Problems in Materialism and Culture*, Verso, London.

Wills J, 1998, 'Building labour institutions to shape the world order? International trade unionism and European Works Councils', Paper presented to the Economic Geography Research Group Seminar 'Institutions and Governance', 3 July 1998, University College, London.

Winterton J, 1985, 'Computerized coal: new technology in the mines', in Beynon H (ed.), *Digging Deeper: Issues in the Miners' Strike*, Verso, London, 231–44.

World Bank, 1994, *Annual Report 1994*, World Bank, Washington, DC.

Wrigley N and Lowe M, 2002, *Reading Retail: A Geographical Perspective on Retailing and Consumption Spaces*, Arnold, London.

Yates C, 1998, 'Defining the fault times: new divisions in the working class', *Capital and Class*, 66, 119–47.

Yearley S, 1995a, 'The transnational politics of the environment', in Anderson J, Brook C and Cochrane A (eds), *A Global World?* Oxford University Press, Oxford, 209–48.

Yearley S, 1995b, 'Dirty connections: transnational pollution', in Allen J and Hamnett C (eds), *A Shrinking World?*, Oxford University Press, Oxford, 143–82.

Yeung H, 1998, 'Capital, state and space: contesting the borderless world', *Transactions of the Institute of British Geographers*, NS, 23, 291–309.

Young G, 1973, *Tourism: Blessing or Blight?*, Penguin, Harmondsworth.

Young J E, 1992, *Mining the Earth*, Worldwatch Institute, Washington, DC.

Zimmerman E, 1951, *World Resources and Industries*, Harper & Row, New York.

Zukin S, 1995, *Cultures of Cities*, Blackwell, Oxford.

Zukin S, 1998, 'Urban lifestyles: diversity and standardisation in spaces of consumption', *Urban Studies*, 35, 825–39.

Index